Modeling Populations of Adaptive Individuals

Modeling Populations
of Adaptive Individuals

STEVEN F. RAILSBACK
BRET C. HARVEY

PRINCETON UNIVERSITY PRESS
Princeton and Oxford

Published by Princeton University Press
41 William Street, Princeton, New Jersey 08540
6 Oxford Street, Woodstock, Oxfordshire OX20 1TR

press.princeton.edu

Library of Congress Cataloging-in-Publication Data
Names: Railsback, Steven F. (Steven Floyd), 1957– author. | Harvey, Bret C., 1958– author.
Title: Modeling populations of adaptive individuals / Steven F. Railsback, Bret C. Harvey.
Description: Princeton : Princeton University Press, [2020] | Series: Monographs in population
    biology; 63 | Includes bibliographical references and index.
Identifiers: LCCN 2019023767 (print) | LCCN 2019023768 (ebook) | ISBN 9780691180496
    (hardback) | ISBN 9780691195285 (paperback) | ISBN 9780691195377 (ebook)
Subjects: LCSH: Population biology—Mathematical models. | Ecology. | Adaptation (Biology)
Classification: LCC QH352 .R35 2020 (print) | LCC QH352 (ebook) | DDC 577.8/8—dc23
LC record available at https://lccn.loc.gov/2019023767
LC ebook record available at https://lccn.loc.gov/2019023768

British Library Cataloging-in-Publication Data is available

Editorial: Alison Kalett, Kristin Zodrow, and Abigail Johnson
Production Editorial: Ali Parrington
Production: Jacqueline Poirier
Publicity: Matthew Taylor and Katie Lewis
Copyeditor: Jennifer McClain

This book has been composed in Times

Printed on acid-free paper. ∞

Printed in the United States of America

10 9 8 7 6 5 4 3 2 1

To Margaret and Maggie, our very adaptive spouses.

Contents

Preface

In 1999 the two of us began overhauling an individual-based model (IBM) whose purpose was to forecast stream trout population responses to novel habitat conditions. We sought to include adaptive behavior of the individual fish because it seemed critical to that purpose. The key problem we faced was how to model habitat selection, the primary adaptive behavior of our model trout. We knew from the literature that trout select habitat as a trade-off between growth and risk: they normally select habitat that supports growth, but they also avoid habitat that puts them at high risk of predation. Further, trout habitat selection depends on the state of the individual and on environmental conditions: fish facing food scarcity or high temperatures that increase metabolic demands may take more risks to obtain food, while fish facing extreme risk of predation may forego feeding to hide for lengthy periods.

Like many others, we initially assumed that this kind of trade-off behavior is so common and fundamental that there must be established ways of modeling it. Instead, we found that very few IBMs included trade-off behaviors and those that did represented them in ways too simple to capture the known adaptive abilities of trout. What we did find very appealing was the *state-based dynamic modeling* theory pioneered by Colin Clark, Alasdair Houston, Marc Mangel, and John McNamara. This theory provides a process for thinking about and modeling trade-off decisions based on the most solid assumption in ecology: that behavior acts to convey individual fitness, including future survival and reproductive output. But like others, we found that this very appealing theory cannot be used directly in IBMs: it is implemented via an optimization that requires assumptions—that future conditions are known and unaffected by individual behavior—that cannot be met when modeling *populations* of adaptive individuals that interact with each other.

To get around the problem that we cannot optimize fitness-seeking behavior in IBMs, we looked for ways to retain the fundamental concepts and dynamics of state-based dynamic modeling without requiring optimization. We arrived at an approach that seemed to suffice for the trout IBM: instead of assuming that an individual optimizes its decisions over a fixed, known future period, we assumed that each individual makes explicit predictions of future conditions and uses approximation to make a good but not necessarily optimal decision, and then updates the predictions and decision routinely as the individual's state and environment change. We published simulation experiments showing that this approach could reproduce a wide range of realistic adaptive behaviors. After long debate, we decided to refer to this approach by the clumsy but descriptive term *state- and prediction-based theory*.

In the meantime, other ecologists were documenting the widespread importance of the kinds of trade-off behaviors that we had struggled to model. Trade-off behaviors and their effects on population and community dynamics (trait-based indirect interactions, nonconsumptive effects, etc.) have been a popular research topic in ecology since the 1990s. Ecologists now widely accept that much of our classical theory is limited by its failure to consider the effects of adaptive behavior.

Another relevant recent trend is the expanding recognition of limits on organisms' decision-making capabilities. Richard Thaler's 2017 Nobel Prize in Economic Sciences for his contributions to behavioral economics indicates the breadth of this trend. Research on cognition and decision-making in the neurosciences is producing clearer and more mechanistic understanding of exactly how animals identify alternatives and select among them. The new field of plant behavior is making rapid progress on similar questions by focusing on chemical signaling. The traditional assumption of optimality in behavioral ecology has been extremely productive, but is increasingly at odds with what we know about how real organisms make decisions. We need alternatives that let us explore how more realistic models of adaptive behavior affect ecology.

Despite these trends, ecology has not yet produced an established way of modeling population and community dynamics that considers adaptive trade-off behaviors. Population-level models become enormously complex but not very practical or general when we try to include trade-off behavior in them. At the individual level, traditional models of behavioral ecology represent individual adaptive decisions but are not useful in IBMs because their simplifying assumptions about the population and environment—usually, that future conditions affecting the individual are known and unaffected by behavior—cannot be met. Instead, we need models of what individuals do that are useful for predicting and understanding what happens at the population level of organization and higher. In principle, we should have relatively simple theory for behavior that, when used in IBMs, produces complex and realistic population and community dynamics. But such theory has not become established in ecology (or in other complex sciences such as economics). In fact, we perceive that many ecologists do not often think across levels of organization and struggle to understand, for example, why classical approaches to behavioral ecology, such as state-based dynamic modeling and game theory, are not useful for predicting population dynamics, or that risk-growth and other trade-offs often need to be addressed not only in evolutionary ecology but also in population and community ecology.

One goal of this book, then, is to introduce the general idea of *across-level theory* and explain what characteristics theory for individual behavior must have if it is to be useful for predicting the dynamics of populations of adaptive individuals. We do this by reviewing existing theory for trade-off behaviors and its usefulness for modeling populations.

Our main goal, though, is to make state- and prediction-based theory (SPT) accessible to ecologists. As far as we know, SPT remains the only kind of theory, based on the same evolutionary concepts as state-based dynamic modeling, that is useful as across-level theory in population and community models. (Like state-based dynamic modeling, SPT is not "a theory" but an approach to developing theory for particular systems.) SPT's departure from the assumption of optimal behavior might be considered a substantial leap, but we think most ecologists will see SPT as an incremental and intuitive change, because it retains the key concept that behavior acts to convey individual fitness while explicitly acknowledging what we all know from our own experience—that decision-making is rarely if ever optimal. SPT assumes that organisms do what we all do: make predictions about an uncertain future, make a decision based on those predictions, and update as things change.

This is, then, primarily a how-to book on using SPT to model individual behavior in IBMs. We hope it will encourage ecologists to extend their understanding of ecological systems through theory that explicitly links individuals and higher levels of organization and to recognize that SPT can provide a framework for integrating population and community ecology with our growing understanding of how animals, plants, and other organisms actually make adaptive decisions. Most importantly, we hope that the book will help ecologists build models that successfully represent the individual trade-off decisions that can be critical to understanding and making useful predictions about ecological systems, particularly those experiencing novel conditions. While we focus mainly (but not exclusively) on single-species models, the methods we present also apply to community and ecosystem models.

Not all readers will want to digest the entire book. Chapters 1 and 2 are motivational. Chapter 1 lays out the problem of understanding and modeling populations of adaptive individuals and reviews the methods ecologists have applied to it. Chapter 2 provides an overview of the model that motivated the development of SPT; a thorough understanding of this chapter is not essential for learning to use SPT, but it does illustrate (along with chapter 7) the very complex and realistic contexts in which SPT can produce successful adaptive behaviors and the population dynamics that result. Readers interested only in learning to use SPT can focus on the introduction in chapter 3 and detailed guidance in chapter 8. However, the example models developed in chapters 4–6 (and their computer code; see below) were designed in part to be modified and adapted to new problems by readers.

We also address how we can develop and test theory for how real systems of adaptive individuals work (chapter 9); how we can establish the credibility of our models (chapter 10); and the benefits of integrated research programs that link modeling, laboratory, and field research (chapter 11). Our final chapter summarizes the lessons learned throughout the book and what those lessons indicate about how ecologists can more fully address the links and feedbacks between individual behavior and the dynamics of populations, communities, and ecosystems.

The models we develop as examples in chapters 4–6 are computational, not mathematical. We hope that readers will take the opportunity to examine, explore, and modify these models, which are accessible at https://press.princeton.edu/books /paperback/9780691195285/modeling-populations-of-adaptive-individuals under Supplementary Materials or at www.railsback-grimm-abm-book.com/MPAI. The models are implemented in an easy-to-understand and well-supported programming language and can be either downloaded or used within a web browser with no additional software.

Acknowledgments

The research and management projects that motivated much of what we present here were funded by institutions including Argonne National Laboratory, the Electric Power Research Institute, Pacific Gas and Electric, Southern California Edison, the US Environmental Protection Agency, the US Fish and Wildlife Service, the US Forest Service, and Western Area Power Administration.

We thank the many collaborators who contributed to this work, including many of our faculty colleagues and graduate students in Humboldt State University's Fisheries Biology, Mathematics, and Wildlife Departments. We also acknowledge Bret's colleagues at the Forest Service's Redwood Sciences Laboratory, especially the many contributions of Rod Nakamoto and Jason White.

In addition to these local collaborators, we are extremely fortunate to interact with and learn from many researchers and modelers around the world. Foremost among these is Volker Grimm, Helmholtz Center for Environmental Research UFZ, Leipzig, Germany; Volker has for many years been our primary source of ideas, feedback, and encouragement. Justin Calabrese (Smithsonian Conservation Biology Institute and University of Maryland) provided extensive advice on analysis of movement tracking data. Øyvind Fiksen (University of Bergen) provided figure 5.1. Other collaborators who have made highly appreciated contributions include Daniel Ayllón Fernández (Universidad Complutense Madrid); Uta Berger (Dresden University of Technology); Jarl Giske, Geir Huse, and colleagues (University of Bergen and Institute of Marine Research, Norway); Marc Mangel (University of California, Santa Cruz); Roger Nisbet (University of California, Santa Barbara); Richard Stillman (Bournemouth University); and many of Volker's colleagues at UFZ Leipzig.

Modeling Populations of Adaptive Individuals

CHAPTER 1

Adaptive Individuals and Population Ecology

1.1 ADAPTIVE TRADE-OFF BEHAVIOR AND ECOLOGY

This book is about explicitly including individual adaptive behavior in the ways we study and model ecological systems. To begin, we must define what we mean by models, systems, individuals, and adaptive behavior.

The *models* we use here follow more in the tradition of the physical sciences and engineering than in the tradition of ecology: we build models with enough detail to be useful for real management problems, and enough theory and mechanism to make useful predictions of system responses under novel conditions of the kind often relevant to managers (e.g., resulting from habitat alteration, introduced species, climate change). This kind of modeling contrasts with the two kinds that many ecologists think of first. When we think of detail and real-world management, many ecologists think first of statistical models carefully fit to data (e.g., Hilborn and Mangel 1997). But statistical models by nature have limited ability to predict responses to novel conditions because such predictions require extrapolation outside the range of conditions used in parameter fitting, and may be unusable anyway because adequate data cannot be collected before model results are needed. When we think of theory and mechanism, we often think of classical theoretical ecology, which uses models often too simplified and abstract for practical applications. In this book we base models on theory and algorithms specific enough to address real systems—and consider the kind of empirical science needed to develop and test our theory and models.

We broadly define ecological *systems* to span populations, communities, and ecosystems. Most of the examples we present involve single-species systems, and often we refer to the systems of interest as populations. However, the methods and techniques we present can also apply to higher levels of organization. The trout model discussed in chapter 2 can be considered a community model in two ways: it can represent multiple competing trout species (e.g., such that stressors on one species may benefit another; Forbes et al. 2019), and it can represent how trout populations affect adjacent trophic levels (e.g., Railsback and Harvey 2011, 2013). Further, the individual-based approaches we consider here have been applied to ecosystem ecology for decades, although their value will no doubt increase with increased use of tested, reusable theory (Grimm et al. 2017).

The *individuals* we refer to are typically individual organisms. However, the individuals in our models might also represent other ecological levels of organization that can be thought of as components of a population or higher-level system. Examples include families of group-living animals, insect colonies, and individual plant stems attached to the same root system. And while this book is explicitly about ecology, all the main ideas and methods we address are also relevant to modeling human systems where the "individuals" could be people, families, governments, businesses, or other organizations.

The individuals we model have *adaptive behavior* in the sense that they make decisions in response to changes in themselves and their environment, presumably to increase their fitness. We focus specifically on adaptive *trade-off* decisions. Many studies and models have addressed behavior with one objective, such as "optimal" foraging that maximizes food intake, but far fewer have addressed populations of individuals that behave to trade off multiple elements of fitness, such as energy acquisition, predator avoidance, and reproductive output. For the sake of conciseness, we often refer to these adaptive trade-off behaviors simply as "adaptive behaviors."

(We recognize that not all individual-based population models need to include trade-off behaviors, and the modeling process is much simpler when trade-offs can or must be ignored. Neglecting trade-offs may be appropriate when modeling taxa or life stages for which individual fitness is dominated by one objective, such as acquiring energy reserves to survive winter, or when we simply do not know enough about what trade-offs are important and how they act. Much of this book remains relevant to systems and problems for which representation of trade-off behaviors is unnecessary or impossible.)

Adaptive trade-off behaviors are common in natural systems and probably often important, as illustrated by the large body of research that has used giving-up food densities to quantify predation risk across a broad range of taxa. Adaptive trade-off behaviors can range from short-term decisions such as habitat and activity selection (deciding when and where to feed or hide), to midterm responses such as energy allocation (e.g., a plant allocating resources to growth or production of pest-inhibiting chemicals), to seasonal decisions such as when and where to migrate, to irreversible life history decisions such as whether to enter a reproductive state. Trade-off behaviors can be quite complex, as in the example of bird molting recently shared with us by our colleague Jared Wolfe. The decision of when to replace feathers can be a complex trade-off involving energetics and growth (replacing feathers is an energy cost but necessary for attaining adulthood), predation risk (escape ability is reduced while feathers are being replaced but enhanced by new feathers once they are in place), and reproduction (adult plumage is necessary for attracting mates but retaining juvenile plumage after maturation can provide males with opportunities for deceptive mating, and molting can affect the ability to care for offspring) (Foster 1987; Mumme 2018).

Understanding and quantifying the importance of adaptive trade-off behavior has been a major theme of ecology in recent decades. The review by Lima and Dill (1990) reinforced the widespread importance of perhaps the most prominent trade-off behavior in animal ecology: the need to forage while avoiding predators. This review drew two conclusions that helped redirect ecology: (1) ignoring predation risk in foraging models, as classical behavioral ecology often did, is unlikely to be realistic; and (2) by reducing foraging, predation risk can have strong effects on individual fitness whether or not an individual is killed. Documenting and quantifying these indirect effects of predation risk then became an important topic of both empirical and theoretical ecology. Abrams (1993) motivated much of this work by pointing out, via a model conceptually similar to the approach in this book, that adaptive behaviors such as predator avoidance can invalidate a key assumption of the classical models of population and community ecology: that the death rate of prey is linearly related to predator density. Werner and Peacor (2003) identified a variety of empirical studies of how adaptive trade-off behavior affects trophic interactions and concluded that the resulting indirect effects are often stronger than the direct effects (e.g., via predation or grazing) represented in classical models. Further studies and reviews (e.g., Miner et al. 2005; Preisser et al. 2005) and the monograph of Schmitz (2010) reinforced the importance of adaptive behavior to the point that its significance at the population and higher levels of organization seems widely accepted. This acceptance extends to the realization that effective management of populations can require explicit consideration of behavior (Caro 2007).

One contribution of this literature is a standardized conceptual model and terminology for one class of examples: the effects of predator avoidance behavior on trophic interactions in a three-level ("tri-trophic") system (e.g., Abrams 1995, 2007; Railsback and Harvey 2013). We present this conceptual model and terminology here because we use it to typify the kind of modeling this book is about: we seek models that can robustly reproduce these trophic interactions because doing so provides evidence of their capability to represent the effects of adaptive behavior on population and community ecology in general.

The tri-trophic model represents a "community" with predator-level members who eat a prey level that in turn consumes members of a resource level. *Consumptive effects* represent actual ingestion: if predators eat 20% of the prey population, then their direct, consumptive effect on prey is 0.2. *Nonconsumptive effects* measure the effects of one trophic level on the next lower level due to behavior: if predator presence causes prey to reduce foraging effort such that they have less energy for reproduction, which results in 30% lower abundance than with unrestrained foraging, then the nonconsumptive effect is 0.3. *Indirect effects* measure trophic interactions across all three levels. *Density-mediated indirect effects* are effects on the resource level due to the consumption of prey by predators: the more prey killed by predators, the higher the resource level. *Trait-mediated indirect effects* include those on the resource level due to the prey's behavioral response to predators: prey reducing their foraging rate to avoid predation also results in higher resource levels. Trait-mediated effects can also be bottom up: changes in resource density can affect prey behavior in ways that affect the rate of prey consumption by predators. For example, higher resource density can have a bottom-up, indirect negative effect on predators by reducing the amount of time prey need to expose themselves to predation while foraging.

These kinds of behavior-mediated dynamics seem straightforward and intuitive in addition to being ecologically important, and much of the relevant literature was published decades ago. We might therefore expect to have established theory and useful techniques for modeling such systems.

1.2 MODELING SYSTEMS OF ADAPTIVE INDIVIDUALS

How has the theory and practice of ecology advanced to incorporate the now-well-known effects of adaptive behavior? We address this question here with a very concise review of several general approaches. To be specific, we address population, community, and ecosystem ecology at subevolutionary time scales, whereas—as this review shows—much of the thinking about adaptive behavior has addressed its evolutionary effects. Our interest is in modeling management and research problems of real systems at ecological time scales, considering adaptive behaviors that yield variable decisions among individuals that are influenced by changing environmental conditions. Of course the main point of this review is to justify the rest of the book by showing that we still need a practical, general approach for thinking about and modeling populations and higher levels of organization that includes the effects of individual adaptive behavior.

System-level models—equation-based models at the population or community level—are the mainstay of classical ecological theory, so it is natural that they have been applied to the ecological effects of adaptive behavior. Abrams (1993) provided an early and particularly influential example: an equation-based model of a population of identical individuals that trade off food intake and predation risk to maximize several alternative measures of fitness that consider survival and expected reproductive output. Other equation-based models by Abrams addressed the tri-trophic model system defined above (Abrams 1995) and habitat choice in a simplified two-species system (Abrams 2007).

Food web models have also been used extensively to examine effects of trade-off behaviors on trophic ecology (Beckerman et al. 2010). These models can, at least conceptually, represent how trade-off behaviors affect population dynamics and trophic relations (static or dynamic) in food webs. An interesting character-istic of at least some of these models is that evolution, instead of or in addition to adaptive behavior, is assumed to be the mechanism that changes traits such as foraging, habitat selection, and life history expression (Loeuille 2010).

System-level models have produced important new ideas and conceptual under-standing of populations of adaptive individuals. Abrams (2010; see also Valdovi-nos et al. 2010), for example, points out that trade-off behaviors (1) can reverse expected outcomes of some trophic interactions, e.g., predation by one species can increase, not decrease, consumption of the same prey by another predator; (2) usually promote stability of communities, but can have destabilizing effects; and (3) often promote coexistence of competing species. However, for addressing specific real ecological problems, system-level models are inherently limited by the mathematical complexity that mounts rapidly with the number of behaviors included, the need to specify the mathematical form of the effects of behavior in advance, and dependence on parameters that are often difficult or impossible to evaluate. Abrams (2010) identifies many challenges of modeling effects of behav-ior in system-level models. These challenges make unsurprising the fact that few if any such models have addressed specific ecological systems and problems, or focused on realistic behaviors for particular sets of species.

Game theory, loosely defined, is a set of approaches for finding optimal or sta-ble solutions for systems of adaptive individuals. Game theory is often assumed relevant to modeling populations of adaptive individuals because it addresses rel-evant processes, such as competition and feedbacks between the system and its individuals. But the game theory most commonly used in ecology is *evolutionary* game theory, which addresses the intergenerational evolution of "strategies" that are, by definition, inherited and fixed during an individual's lifetime. Game theory has proved essential in understanding evolutionary ecology (Brown 2016).

Game theory has also been applied to the challenge we address: representing how individuals make day-to-day adaptive decisions. Applications include trade-off decisions such as how individuals select among habitats that differ in food intake and predation risk, with feedbacks from the behavior of other individuals (e.g., Alonzo 2002; Alonzo et al. 2003). These applications use the basic concep-tual framework of behavioral ecology—that behavior acts to maximize survival or reproductive output over a time period—and then apply iterative games to solve the state of a system of adaptive individuals.

However, the use of game theory to model systems of adaptive individuals has both practical and conceptual drawbacks. Practically, the mathematical and computational challenges of setting up and solving games can become overwhelming as realistic complexities such as spatial and temporal variation in habitat, variation among individuals, and large numbers of decision alternatives are added. Conceptually, nonevolutionary game theory uses iteration to optimize each individual's behavior as if the individuals all know and respond to the state of the entire system and the entire system immediately responds to the behavior of its individuals. While these assumptions may be useful for a general understanding of the system, they ignore realistic limitations on individual perception and adaptation (e.g., local instead of global information and interaction) and how those limitations can affect a system.

Adaptive dynamics has several times been mentioned to us as relevant to modeling systems of adaptive individuals. This term refers to a conceptual and mathematical modeling approach to predicting system responses to trait variation among individuals (Geritz et al. 1998). However, this approach is not relevant to the problems we address, for the same reasons that many game theory applications are not. First, adaptive dynamics addresses how populations and communities evolve—how the frequencies of specific traits change over generations—not how system characteristics such as abundance or persistence depend on individuals over time scales relevant to management decisions. Second, it treats traits as static, inherited characteristics of individuals instead of as the dynamic behaviors that we are interested in. Finally, the mathematics become very difficult to solve unless the modeled system is highly simplified. While this approach may be useful for thinking about evolution in less-simplified ways, it does not appear useful for making testable predictions of real ecological systems.

Individual-based models (IBMs; also often called "agent-based" models; Vincenot 2018), used in this book, model populations or communities by explicitly representing the individuals that make up the system, and so alleviate the need for system-level theory and parameters that are major obstacles to modeling effects of adaptive behavior. Further, IBMs readily represent other important complexities, such as variation among individuals, full life cycles, and spatial and temporal variation in habitat conditions. (For in-depth information on IBMs, see Grimm and Railsback 2005 and Railsback and Grimm 2019.) In IBMs of adaptive individuals, system dynamics emerge from the outcomes of interdependent individual trade-off decisions.

Another important characteristic of IBMs is that they naturally represent the feedbacks between individual and population levels. In a simulated system of interacting, competing individuals, the state of the system depends on the behavior of its individuals, and the behavior of the individuals depends on the state (at least locally) of the system. Population-level characteristics such as density, age structure, and spatial distribution emerge from individual adaptive decisions, while each individual's adaptive decisions can depend on the number, ages, and sizes of other individuals it competes or cooperates with.

While the bottom-up nature of IBMs contributes to the belief that IBMs are a theory-free approach that relies on simulation alone, our experience and the scientific literature indicate that we do indeed need theory for trade-off decisions in IBMs. It turns out not to be trivial to model individual trade-off behaviors in ways that cause realistic population and community dynamics to emerge, and few IBMs do so. We anticipate that IBMs will be more productive when based on general, reusable, and well-tested theory (Grimm 1999).

1.3 ADAPTIVE BEHAVIOR IN INDIVIDUAL-BASED MODELS

If we use IBMs to model populations of adaptive individuals, how do we represent the adaptive behavior? The behavioral ecology literature is a natural place to look for solutions, but our attempts to apply models from that literature to IBMs have not been fruitful. In a recent review of behavioral ecology in marine science, Dill (2017) similarly concluded that, despite its clear importance, there has been little progress incorporating adaptive behavior in population or community models for conservation. How can it be hard to represent individual adaptive behavior (and not just trade-off decisions) in IBMs, when behavioral ecologists have been working on this problem since the pioneering foraging models of Emlen (1966) and MacArthur and Pianka (1966)? In our estimation, models of adaptive behavior that have proved extremely useful in other contexts are not useful in IBMs (which is of course not surprising because they were not designed for that purpose).

Much of what we think of as classic behavior theory was developed by using simplifying assumptions to reduce problems until they have straightforward mathematical solutions. An example that appeared often in the early IBM literature is the μ/G rule: the theory that trade-offs between risk and feeding can be optimized by selecting the alternative that minimizes the ratio of risk to growth. This theory describes optimal decision-making under very specific conditions (e.g., Gilliam and Fraser 1987; Clark and Mangel 2000); its assumptions are rarely met in general (Tyler and Rose 1994). Our attempt to use this rule in an IBM of stream trout (see chapter 2) failed in several ways (Railsback et al. 1999). Werner and Anholt (1993) and Mangel (1994) also point out problems with applying the μ/G rule outside the conditions it was specifically derived for.

Dynamic state variable modeling, or DSVM (Mangel and Clark 1986, 1988; Houston and McNamara 1999; Clark and Mangel 2000), was a major advance in behavior theory. DSVM involves defining a measure of the individual's fitness at a future time, then finding the sequence of behaviors that maximizes that expected future fitness. DSVM appears highly appealing for IBMs because it provides a conceptual framework and tools (usually, dynamic programming optimization) for making trade-offs, e.g., between foraging and avoiding predation. However, application of DSVM to IBMs has proved problematic (e.g., Luttbeg et al. 2003). While the conceptual framework of DSVM is extremely flexible and powerful (we use it throughout this book), finding optimal decisions requires assumptions that cannot be met in IBMs of even minimal complexity. Most importantly, finding optimal decisions requires individuals to know future conditions, such as the availability of food and perhaps predator avoidance habitat, whereas in an IBM the future availability of such resources is essentially unknowable because it depends on the behavior of all the other individuals. In other words, this classic theory of behavioral ecology does not accommodate the feedbacks of behavior inherent to IBMs and real ecological systems.

So how have individual-based modelers represented adaptive trade-off behaviors when the theory of behavioral ecology has not been useful? We are aware of four general approaches, excluding the one covered in the rest of this book. Grimm and Railsback (2005) reviewed these approaches; we summarize them here.

First, some modelers have represented trade-offs using simple ad hoc rules. Ward et al. (2000) modeled a system resembling the tri-trophic model by assuming that foragers behave to obtain an adequate food intake, while predation risk was represented simply as a reduction in the rate of food intake due to time spent on vigilance instead of feeding. The μ/G rule discussed above has also been used to model trade-off decisions in IBMs, even though its limitations, such as not producing useful decisions when growth is negative, required further ad hoc modifications (e.g., Van Winkle et al. 1998). This approach appears to be most useful when adaptive behaviors can be highly simplified, with the loss of realism an acceptable cost of simplifying the model.

Second, many IBMs represent adaptive behavior not with theory but with stochastic rules based on empirical observations. This approach treats observed rates at which different decision alternatives were chosen by real individuals as probabilities in stochastic decision models, reflecting the classical population modeling tradition of fitting models to field data. Gusset et al. (2009) provide a simple example that models many behaviors of wild dogs as stochastic decisions with probabilities estimated from extensive field observations. This approach is undergoing a technological explosion, with tracking devices and other sensors increasing in capability and decreasing in cost while techniques for fitting statistical models of behavior to the sensor data increase in sophistication. These technologies are of obvious value for understanding and quantifying behaviors such as foraging decisions and selecting home range sizes and locations. We discuss these technologies and their potential, including potential contributions to theoretical understanding, in section 11.4.

A third approach to modeling adaptive trade-off behavior is detailed simulation of the events driving the behavior. Schmitz's tri-trophic model of a spider-grasshopper-plant system (Schmitz 2000, 2001) provides an important and successful example of this approach. This model represents how grasshoppers adapt their choice of which plant to eat as a trade-off between resource intake from feeding and the risk of predation by spiders: one plant species provides more nutrition but more exposure to predation than a second plant. The behavior was modeled via detailed representation, over short time steps, of how each spider moves and how grasshoppers detect spiders and move among plants in response. Like the previous two, this approach is not theoretical: it substitutes detailed representation of a specific system for decision-making theory. Its applicability is to models in which adaptive behavior can be broken down into a series of responses and actions instead of being treated as a decision-making process.

The fourth approach allows models of adaptive behaviors to evolve in computer simulations. Genetic algorithms and artificial neural networks have long been applied to complex problems in many fields, with great success. This approach models a behavior using a mathematical representation of a simple neural system, with inputs for the information used in the decision and outputs signaling the resulting decision. The complex links between information inputs and decision outputs are determined by parameters, and good parameter values must be found via artificial evolution. This approach requires an IBM in which the behavior parameters can evolve, which means the IBM must represent a population of individuals that each have different parameter sets, survival and reproduction based on the fitness provided by the behavior model parameters, and reproduction with processes resembling genetic crossover and mutation to generate new parameter sets. This approach has been used to evolve models of adaptive trade-off behavior in a variety of IBMs, most prominently by Jarl Giske and his research group at the University of Bergen, Norway (e.g., Huse and Giske 1998; Huse et al. 1999; Strand et al. 2002; Eliassen et al. 2016; but see also Morales et al. 2005). Peacor et al. (2007) used evolved adaptive behaviors in an individual-based "virtual ecosystem" designed to study exactly the kind of problem we address in this book: how predator avoidance behavior affects population and community dynamics. An extensive literature addresses the use of genetic algorithms and artificial neural networks for solving complex problems, mainly in the computer science and engineering literature; Mitchell and Taylor (1999) provide an overview for biologists. We generally characterize this approach as empirical instead of theoretical because the behavior model parameters are fit to be successful in the artificial systems in which they evolved, and because artificially evolved neural networks can be very difficult to interpret into rules or equations that we can understand and generalize from. However, recent work by Giske et al. (2013, 2014; Budaev et al. 2018) attempts to tie artificially evolved behavior models to mechanisms, such as responses to competing emotions, by which real organisms make decisions. This work has the potential to link the behavior models used in IBMs to the neural mechanisms of real organisms and thereby contribute to theoretical understanding of adaptive behavior.

While these four approaches have been successful in some ways—for example, the models of Ward et al. (2000) and Schmitz (2000) succeeded in producing the indirect interactions characterizing the model tri-trophic system defined in section 1.1—they share common problems that were motivation for the approach we first introduce in section 1.6. None are theoretical, which means that the behavior models rely on system-specific empirical assumptions, data, or artificial evolution. Dependence on empirical information can make these approaches good at reproducing observed behavior, but also makes them less useful for the important goal of predicting system responses to novel conditions, as in assessing potential effects of habitat alteration or climate change. Their empirical nature also means that new models must be built for each system and behavior, instead of reusing general theory. In chapter 9 we present a practical process for testing the models of behavior we use in IBMs, by hypothesizing alternative models and falsifying them in simulation experiments; this process could document and improve the usefulness of any of these approaches. Testing and comparing alternative ways of representing behavior in IBMs is becoming more common but has been rare.

1.4 ADAPTIVE BEHAVIOR, PHYSIOLOGY, AND NEUROBIOLOGY

We are certainly not experts in physiology or neurobiology, but it is clear that rapid advances in these fields are revealing more and more about how real organisms actually make decisions. If our goal is to model real systems that depend on adaptive behavior, then we need to have some understanding of what is being learned in these fields and the capability of our theory to represent actual mechanisms of adaptive behavior.

The importance of physiology to adaptive behavior has long been clear. For example, the "state" in the DSVM approach usually includes a physiological variable representing energy reserves. Energetic state is a critical link between behavior and individual fitness: behavior determines how much energy an individual acquires and spends, and physiology determines how the resulting energy balance affects survival of starvation, disease, and reproduction. Therefore, modeling behavior as we do in this book requires some ability to model energetics and its effect on survival and reproduction. At some point our ability to model behavior can be limited by our ability to model physiology.

Recent research has explicitly addressed the physiological mechanisms through which organisms actually make decisions. The resulting knowledge is relevant to this book in two ways. First, it provides evidence of the potential to replace traditional assumptions of "optimal" decision-making based on perfect information with more realistic assumptions. Second, physiological research can provide direction to us as we look for more useful and realistic models of behavior. We do not believe that ecologists ever posed or used optimal foraging theory because they believed animals are capable of perfect decisions, but instead because the optimality assumptions made it easy to explore good decisions and their consequences. But now we may be able to define and explore more realistic assumptions—or perhaps show when optimality assumptions are not so bad.

In a summary of what neurobiology has contributed so far to the understanding of decision-making in animals, Glimcher (2016) wrote:

> Nearly all current neurobiological evidence suggests that we can think of mammalian (including humans) decision making as being organized around three basic processes:
>
> 1. A suite of brain areas that construct an estimate of the value of each of the options being considered by the decision maker
> 2. A smaller overlapping suite of brain areas that actually compares these option values mechanistically and passes the resulting choice to movement control (or output) systems of the brain
> 3. A set of (again overlapping) learning-related areas that compares the quality of the obtained option with the expected quality of that option and updates the internal representation of option value to improve future decision making

Clark (2013) described a brain as "bundles of cells that support perception and action by constantly attempting to match incoming sensory inputs with top-down expectations or predictions," essentially a "prediction machine."

These descriptions of what actually happens in brains indicate that a behavior-modeling paradigm that assumes individuals identify alternatives and evaluate them using prediction—such as the one we describe in this book, but unlike approaches that identify a single "optimal" choice—may be physiologically realistic and, therefore, more likely to reproduce novel detailed observations of behavior. But this understanding also indicates that prediction and learning, which are generally neglected in IBMs and in much of behavioral ecology, may be more important than we realize.

However, we must remember that this book is about ecological modeling, and we are ultimately interested in dynamics of populations and communities, not just individual behavior. As we understand in more detail how real organisms (and other kinds of individuals) make decisions, we must carefully determine how much of that detail to include in our models: IBMs should, ideally, contain just enough "essence" of real decision-making to let us solve population-level problems. Our goal is not to produce "realistic" models, but useful ones. For example, ignoring learning and instead assuming individuals have competent behavior from birth is often perfectly appropriate for modeling populations. How to decide when a model of behavior is good enough for modeling populations is of course a critical question, which we address in chapter 9.

1.5 WHAT WE NEED TO LINK BEHAVIORAL AND POPULATION ECOLOGY: ACROSS-LEVEL THEORY

Why is this book, which is mainly about modeling individual behavior, in a series on population biology? The previous sections provide only a brief and incomplete review of the thousands of models and publications that have addressed the problem of modeling populations and communities of adaptive individuals, but they are intended only to illustrate one key lesson that we (and others; e.g., Stillman et al. 2015) have learned in 20 years of work on this problem: that IBMs are a practical and productive way to model real systems of adaptive individuals with sufficient complexity and realism to understand and predict their dynamics in useful ways. But for IBMs to be practical and productive, we often need theory for adaptive trade-off decisions, and the kind of theory we need is not yet widely established.

We cannot model populations of adaptive individuals by looking at just the population level *or* just the individual level. Traditional theory in population and community ecology includes system-level equations, which are clumsy at best for representing the complex effects of individual behavior. But the traditional theory of behavioral ecology does not provide what we need for IBMs either, because it neglects the feedbacks from the system to the individual. Behavior theory such as DSVM is very powerful and useful, but only for modeling what an individual does when its behavior does not affect the system it occupies—this theory cannot find "optimal" behavior unless the system is fixed and known.

What we need, then, is *across-level* theory—theory that simultaneously considers both individual and higher levels of ecological organization. Game theory is one kind of across-level theory, as it considers how a whole system of individuals should behave. But, as we discussed above, game theory is practical only for modeling relatively simple situations.

For IBMs capable of addressing most real ecological problems, the across-level theory we need represents individual behavior in ways that are useful for modeling population and higher-level dynamics while including the feedbacks of higher levels on individuals. This book, therefore, addresses how to model individual adaptive behavior when the system is not fixed and known but instead when each individual must make decisions that depend on the decisions of many other individuals.

The across-level behavior theory we need cannot require optimization because optimization is not feasible, or realistic, in populations of adaptive individuals. Giving up optimization may make mathematically inclined ecologists uncomfortable because optimization theory is well-defined, established, and productive. But giving up optimization not only lets us address feedbacks, it also opens the door to approaches that better accommodate the rapidly growing knowledge of how organisms and other kinds of individuals actually make decisions, including decisions limited by cognitive and perceptual constraints. Optimization is a very powerful way of thinking about many ecological problems, but as we try to include more of nature's complexities in our models, we find limits to its usefulness.

1.6 STATE- AND PREDICTION-BASED THEORY (SPT)

We did not set out as theoretical ecologists seeking to address all the issues outlined above. Instead, we are applied ecologists who started out trying to solve one particular problem with the trout management model described in chapter 2. But we quickly learned that ecology did not offer theory—i.e., general models useful for understanding and predicting real systems—for modeling populations of individuals that use adaptive trade-off behaviors. We needed theory for how an individual makes trade-offs between fitness elements like growth, survival, and reproduction, in IBMs that contain enough complexity to understand real ecological systems or solve typical management problems. These complexities include the following:

- Individuals are unique in characteristics such as size, energy reserves, life stage, and social status.
- Models must represent full life cycles and therefore the ways that organisms change throughout their lives.

- Individuals interact with their environment and each other locally, so the success of their behavior can be constrained by limited knowledge of their habitat and population.
- Individuals interact with each other in ways that produce feedbacks of behavior: the range of options available to each individual, and their payoff in growth and survival, depend on the behavior of others.
- The environment varies spatially to realistic extents; often much of the available habitat provides negative growth or high risk.
- The environment changes over time in at least partially unpredictable ways.

We discovered that we could produce successful trade-off behaviors in IBMs with these complexities by keeping a basic concept of behavioral ecology—that individuals seek to maximize a probabilistic measure of future fitness that combines survival of predation and starvation—while backing off the assumption that this measure is optimized at either the individual level by dynamic programming (as in DSVM) or the system level via game theory. Instead, we assume individuals use a familiar approach to decision-making: make an explicit prediction about future conditions, use approximation to find a good solution, and update the prediction and decision routinely as conditions change.

After deciding that this approach was not just a trick for modeling one behavior of trout but instead of general applicability in ecology, we gave it the name *state- and prediction-based theory* (SPT). (Like game theory or DSVM, SPT is not one theory but a set of concepts for modeling specific systems and problems.) What sets SPT apart from its predecessors is that individuals base good, but non-optimal, decisions on explicit predictions as well as on their current state. We also discovered that this approach makes it straightforward to model the population-level effects of realistic constraints on adaptive behavior.

To us, SPT seems a natural way to make the fitness-seeking concept of behavioral ecology useful in models with feedbacks of behavior. We have found similar approaches developed independently in ecology (e.g., Luttbeg et al. 2003) and in economics (e.g., Riddle et al. 2015), where it is a natural adaptation of established decision analysis methods such as cost-benefit analysis. But as far as we know, others have not tried to formalize the approach or explicitly discuss its general use.

1.7 MONOGRAPH OBJECTIVES AND OVERVIEW

This book is our effort to describe SPT as an approach to modeling systems of adaptive individuals, using examples and general guidance, and to consider how it can be integrated with empirical ecology as a comprehensive framework for population and community ecology. We explore how good the decisions produced by SPT are by comparing them to optimal decisions in relatively simple situations. But we mainly focus on more realistic, complex situations where optimization is not feasible. We also address the use of SPT as across-level theory. Even though the topic has been covered elsewhere (e.g., Grimm and Railsback 2005, 2012; Grimm et al. 2005), we demonstrate how we can test SPT against observations of real systems and show that it reproduces key patterns at both individual and system levels. Integration of individual-level biology with population ecology is addressed throughout the book.

Many themes of this book depart from the traditions of theoretical ecology. First and foremost, the starting place for developing theory is a real-world ecological management problem that needs to be solved, not "first principles" or a particular mathematical framework. Instead of simplifying our study systems to allow use of purely mathematical approaches, we find tools—simulation, approximation, and math—that work in realistically complex situations. Second, we provide a practical process for testing theory against a variety of observations that do not necessarily include long time series of population data. Third, we do not assume that behavior is optimal but instead that it has evolved to be adaptive in the complex, unpredictable world that organisms live in. These characteristics may make those who equate theory with mathematical rigor uncomfortable. However, we hope to encourage testing and refinement of theory for individual adaptive behavior by how well it explains population and community ecology, and to ultimately encourage development of powerful ecological models to address real-world problems.

The natural place for us to start, in chapter 2, is by introducing the model and research program that first set us on this path. While this chapter is not essential for learning to use SPT, it provides the motivation for it and a realistic example. Next, in chapter 3, we provide an overview of SPT and five major steps in using it. The following four chapters provide example applications of SPT to models of increasing complexity. Each of these examples compares the results of SPT to more traditional optimization-based theory; direct comparison is possible for the simplest models, and for the more complex examples we compare general patterns in results.

Chapters 8 through 11 provide our advice and instruction on the use of SPT to develop and test population and community models. Chapter 8 provides detailed guidance for using SPT, organized by the five major steps of developing and testing an IBM that includes adaptive individual behavior. Chapter 9 then provides methods for testing and refining SPT—how to use a hypothesis-testing cycle to pose alternative models of adaptive behavior, test the alternative models against observations, and define the specific contexts under which theory is useful. We then address model credibility in chapter 10, which is largely based on our experience validating and supporting population-level IBMs with field and laboratory studies. Chapter 11 provides broader discussion of how empirical research interacts with SPT. There we look at how modeling populations of adaptive individuals can benefit empirical research programs and identify the kinds of research most likely to benefit modeling. Finally, we wrap up by summarizing the key characteristics of the methods we describe here and listing the lessons learned from all the models examined in the book.

CHAPTER 2

Case Study: Modeling Trout Population Response to River Management

2.1 INTRODUCTION AND MODEL PURPOSE

Our experience modeling populations of adaptive individuals began with building a model for predicting how river and watershed management affects communities of salmonid fishes. (We have actually produced a family of models customized to specific purposes. We refer to them in general using the name "inSTREAM" or simply as "the trout model.") This chapter describes the key characteristics of inSTREAM and how it represents adaptive trade-off decisions, and provides the background needed to understand its design and complexity.

The chapter has several purposes. First is simply to provide background on our own experience and illustrate how working on ecological management problems can quickly lead to the need for models that cope with adaptive individual behavior. Second is to introduce and illustrate the methods that we develop further in the rest of this book; readers interested only in how to use SPT could skip ahead to chapter 3. However, the third purpose of this case study is to illustrate how much natural complexity can be included in an ecological model while still producing usefully realistic individual behavior, and the system dynamics that emerge from behavior.

The initial purpose of inSTREAM was to assess how alternative reservoir operation rules, which produce different patterns of flow and temperature in downstream waters, affect populations of sympatric trout species. It quickly became apparent that such a model would also be useful for a variety of management applications and for exploring more general ecological questions.

A common theme among complex IBMs is that many were developed to displace simpler modeling approaches that have proved poorly suited, yet persistently popular, for supporting management decisions (Stillman et al. 2015). The river management questions our trout model was designed for have traditionally been addressed with simple habitat selection models. Habitat selection models in particular have major weaknesses, especially in not being able to address variability over time in flow and temperature or the cumulative effects of flow, temperature, and biological interactions. These models also cannot make predictions that are either testable or directly relevant to management decisions (Railsback 2016). The idea that IBMs could provide a better alternative to traditional river management models originated with our former colleagues at Oak Ridge National Laboratory, who developed the first prototypes (Jager et al. 1993; Van Winkle et al. 1998).

Early in model development it became clear that for an IBM to be useful for management applications and to avoid the limitations of traditional approaches, it needs to be quite complex. Stream habitat is innately complex and variable over both space and time (figure 2.1). Flow and temperature can vary dramatically at time scales from less than a day (due to, e.g., hydropower operations or storms) to years (e.g., due to interannual weather variability or climate change), which means that the model must both operate at a short time step and simulate many years, and, therefore, simulate full life cycles. Both flow and temperature have a variety of effects on individual growth, survival of predation and other risks, and reproductive success; yet other factors such as food availability and

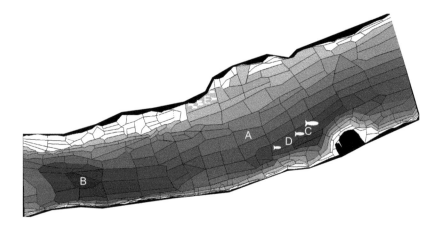

FIGURE 2.1. Example habitat relations represented in the trout IBM. Habitat is modeled as polygonal cells in the horizontal plane, seen here in a top-down view with shading representing depth. Adult trout growth is highest in cells of moderate velocity such as (A), whereas predation risk, primarily from birds, is lowest in deep cells (B). Cells offering the best trade-off provide positive growth and low predation risk, such as (C). However, competition for food in a size-based hierarchy means that the best cell may support only a few adult trout, with smaller individuals forced to use other cells (D). Very small juveniles, however, are at greatest risk from adult trout and grow best at lower velocities; hence they select shallow habitat that adults avoid (E). Because each juvenile consumes relatively little food, they can feed profitably at higher densities than adults can.

the intensity of several kinds of predation also affect these elements of individual fitness. Further, intra- and interspecific competition have clear and strong effects: multiple salmonid species compete for the same food within the same habitat and commonly establish dominance hierarchies driven by body size, with dominant individuals excluding others from desirable locations (Hughes 1992).

Because of these complexities, the trout IBM needed to include a variety of individual-level processes, such as feeding, bioenergetics, various causes of mortality, and spawning. Fortunately, trout have been the subjects of extensive research for many years and existing literature covered many of these key processes. In streams, trout commonly feed by holding position and making short forays to capture food as it drifts by in the water column, so food intake and the energy cost of feeding depend on local hydraulic conditions. A variety of hydraulic models are available to predict local water depth and velocity in microhabitat patches (or "cells," at approximately the scale that trout select feeding habitat), as a function of river flow. Feeding models (e.g., Hughes and Dill 1990) predict how trout food intake varies with microhabitat depth and velocity, and laboratory studies have shown how feeding is also affected by turbidity (e.g., Barrett et al. 1992). Bioenergetics models (e.g., Hanson et al. 1997) predict trout growth from food intake, swimming speed, and temperature.

2.2 ADAPTIVE BEHAVIOR IN THE TROUT MODEL: HABITAT SELECTION

Despite the existence of models for many components of the trout IBM, the goal to include adaptive behavior presented a major unresolved challenge. Habitat selection is widely accepted as a key mechanism relating habitat conditions to population dynamics in salmonids: fish respond to changes in flow and sometimes temperature, and changes in their own state variables such as size and energy reserves, by selecting different microhabitats. While early habitat selection models assumed that this behavior optimized growth (e.g., Hughes 1992; Hill and Grossman 1993), avoidance of predation risk can also be an important factor driving trout habitat selection. While some locations may offer both adequate growth and relative safety from predation, such locations may not always be available, especially for subdominant individuals. For any individual, the microhabitat offering highest growth is unlikely to also offer lowest predation risk, so a trade-off must be made. This trade-off may be most apparent in the behavior of juveniles: small trout are characteristically found in shallow habitat, presumably because it is safest from predation by larger fish, which avoid shallows because they expose large fish to high risk of predation by birds. The trade-off is also influenced by individual condition: fish with depleted energy reserves need higher energy intake even if it requires greater risks.

The challenge of representing this adaptive trade-off behavior by trout in population-level models well reflects issues raised in chapter 1. We recognized that DSVM provides the right conceptual framework for modeling this decision, but in realistic IBMs it could not be used in its standard way. Successful representation of habitat selection by trout clearly requires incorporation of intraspecific interactions: resources available to individuals, and therefore their habitat selection decisions, depend in part on the number, size distribution, and behavior of other individuals. Even without these population feedbacks on behavior, the complexity of the behavior would preclude use of DSVM. Useful simulations of habitat in the trout model can span decades and include thousands of habitat cells that vary unpredictably through time in key physical characteristics such as water velocity, depth, temperature, and turbidity. Using optimization would require us to assume that individual trout know these characteristics of many cells for months or years into the future, and solving the optimization for even one trout would be practically impossible. This environmental complexity in combination with interactions among individuals led us away from approaches that seek to identify optimal sequences of decisions.

Previously explored approaches to habitat selection in stream fish IBMs did not offer formulations that captured all the elements of adaptive trade-off behavior. In the IBM of Clark and Rose (1997), fish considered only growth rate during habitat selection. Van Winkle et al. (1998) developed a functional ad hoc application of the "minimize μ/G rule" in a stream salmonid IBM, but it appeared to overvalue modest increases in growth potential at the expense of greater predation risk and also did not capture the changing value of growth with changes in fish body size, condition, or life stage. The approach of Van Winkle et al. (1998) required the addition of a constant to the denominator of the fitness measure to cope with the common occurrence of habitat providing negative growth. The minimum possible value of the constant would depend on various specific features of particular simulations, while the specific value of the constant of course influences the growth versus risk trade-off.

Given the limitations of these existing approaches, we sought to adapt the conceptual framework of DSVM to the realities of our model. The results were the adaptive behavior theory we now call SPT and the trout IBM now called inSTREAM. The habitat selection approach we developed (Railsback et al. 1999; Railsback and Harvey 2002; also discussed in more detail in chapter 8) uses as its basic fitness measure the expected survival of both predation and starvation until a "sliding" time horizon that is a constant number of days in the future. To represent the fitness benefit to juveniles of growing to adult size, this expected survival is multiplied by a second term: the fraction of adult size the individual is predicted to be at the time horizon. On each simulated day, each individual trout selects the habitat cell that offers the highest value of the fitness measure, assuming the fish stays in the cell over that time horizon. Competition and dominance hierarchies are represented by executing habitat selection from the largest to the smallest fish, with each fish in a cell reducing the food available to smaller individuals. The individuals predict conditions affecting this fitness measure—their predation risk and growth rate—by simply assuming that current conditions will persist over the time horizon. Individuals are almost always wrong in assuming that they will stay in one cell over the time horizon with growth and risk remaining constant, yet these assumptions allow individuals to make good adaptive decisions.

Expected survival of predation to the time horizon is modeled as $S_T = S_t^{T-t}$, where S_t is the probability of surviving predation in the current day t and T is the time horizon. With inSTREAM's sliding time horizon, $T - t$ is constant and commonly set to 90 days. For stream fish, we distinguish two sources of predation risk: birds and mammals versus other fish. We make this distinction because the effectiveness of these two kinds of predators varies with habitat. For example, birds and mammals are usually more effective in shallower water, while shallow water can provide a refuge from predation by fish. For both sources of predation, the model includes a parameter representing survival probability in the most risky habitat. Habitat features can reduce that risk, with the reduction represented via logistic relationships (figure 2.2). The risk of predation from birds and mammals can be influenced by water depth, water velocity, distance to cover, turbidity, fish size, and the amount of time spent feeding. The risk of predation from other fish can be influenced by water depth, turbidity, fish size, the amount of time spent feeding, temperature, and predator density.

Unlike many DSVM models (discussed in sections 4.5 and 8.4.3), inSTREAM represents expected survival of starvation over the time horizon as a continuous function that can approach but not reach zero as reserves decline (Railsback et al. 1999). We model daily probability of surviving starvation S_t as a logistic function of "condition" K, the individual's weight divided by the weight of a healthy individual of the same length. When K is 1.0, S_t is close to 1.0, but S_t declines steeply as K falls below about 0.8. Fish generally change condition slowly; this survival model reproduces their ability to survive long periods of weight loss, but also their increasing vulnerability to disease as condition declines. Expected starvation survival over the time horizon S_T depends on the current value of K and its trajectory over the time horizon, which depends on growth. The value of S_T could be calculated by predicting K and the corresponding S for each day to the time horizon and multiplying all the values of S together, but we found that the first moment of S between t and the time horizon T provides a computationally

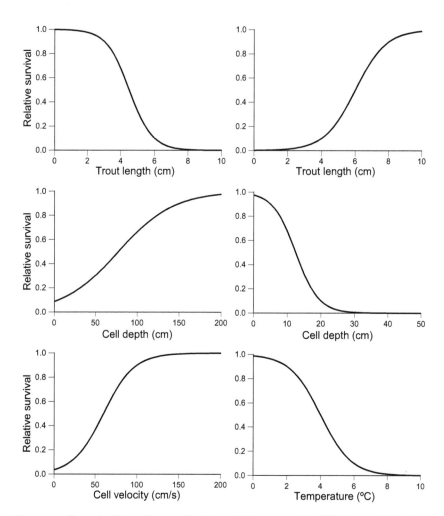

FIGURE 2.2. Example effects of trout and habitat variables on survival of (left) terrestrial predators and (right) piscivorous fish assumed in inSTREAM. The Y axis is the fraction by which daily survival probability is increased by the variable on the X axis. For example, the effect of fish length on survival of terrestrial predators (top left panel) is 0.9 when length is 3.0 cm, so daily survival probability is increased by 90% of the difference between the worst-case survival probability and 1.0. Top: Vulnerability to terrestrial predators increases with fish size because larger fish are both more visible and more valuable to such predators (left), whereas increasing size reduces vulnerability to gape-limited fish predators (right). Middle: Increasing depth makes fish harder to see and catch from the surface, reducing vulnerability to terrestrial predators (left); because shallow habitat is risky for large fish, it conveys safety from fish predators (right). Bottom left: Increasing velocity also decreases visibility and vulnerability to terrestrial predators, but is assumed not to convey safety from fish predators. Bottom right: Fish predators are assumed less active at cold temperatures, which reduces their metabolic demands and feeding rate; in contrast, most terrestrial predators are warm-blooded and do not reduce feeding in cold weather (but temperature could affect terrestrial predation where predators hibernate or migrate, or ice cover occurs).

efficient good approximation (Railsback et al. 1999). (Section 8.4.3 provides a more detailed explanation of modeling starvation survival.)

We chose a sliding time horizon (section 8.3) instead of a terminal date because the trout IBM has no imposed end to the adult life stage. The commonly used time horizon of 90 days is a compromise between two considerations. First, the further the time horizon, the better it approximates "lifetime" fitness, the need to balance energy intake vs. predation risk over the remaining life span. Such a long time horizon is necessary to represent lifetime fitness in part because trout can survive without food for months (Simpkins et al. 2003). But second, the prediction by fish that current conditions persist over the time horizon becomes less reasonable as the time horizon lengthens. Relatively long horizons in variable environments yield decisions that overemphasize predation risk under favorable feeding conditions and underemphasize risk when feeding conditions are poor.

This formulation of SPT provides the model trout with complex and realistic habitat selection responses to a wide range of conditions. We demonstrated this with a series of simulation experiments (Railsback and Harvey 2002). During an unforeseeable physical challenge (a flood), individuals immediately moved to safe and productive habitat on stream margins and then returned to original positions as the flood receded. Competition from larger fish caused shifts to less productive habitat, and increased predation risk caused shifts to safer habitat. Temperature also affected habitat selection: at lower temperatures, metabolic rates are lower so less food is needed to maintain the same growth, so simulated fish shifted to safer habitat. A sudden reduction in food availability caused fish to immediately shift to higher-intake, higher-risk habitat to reduce the risk of future starvation. Use of the model has also illustrated how realistic adaptive behavior is likely to be important in environmental assessments. For example, the model's inclusion of the effects of turbidity on both foraging efficiency and predation risk led fish using SPT to realistically alter their habitat selection under different turbidity regimes (Harvey and Railsback 2009).

InSTREAM also produces common population-level patterns observed in salmonid fishes (Railsback et al. 2002). These include (1) "self-thinning," a negative power relation between mean individual biomass and abundance; (2) a "critical period" of intense density-dependent mortality in young-of-the-year; (3) high age-specific, interannual variability in abundance under certain natural conditions; (4) density dependence in growth; and (5) fewer large trout when pool habitat is reduced.

The power and flexibility of SPT for modeling trade-off behaviors is illustrated by a modification of inSTREAM into a river management model for salmon (Railsback et al. 2013). Adult salmon, upon returning from the ocean to spawning streams, do not feed and instead select habitat that provides protection from predators and low swimming costs (to avoid consumption of energy that would otherwise be used in reproduction). Reproducing this habitat selection behavior required only one extremely simple modification of the model and no change to the habitat selection theory: we simply replaced the feeding submodel with the assumption that food intake is zero for adult salmon. With this change, the same habitat selection theory that maximizes survival of predation and starvation until a future time horizon now causes fish to select safe habitat with low energy costs, because survival of starvation is now driven entirely by the metabolic costs of swimming. Without the motivation to feed, simulated salmon expend much less energy and experience much less predation mortality than do model trout, whose habitat selection trade-off includes the benefits of feeding.

2.3 A SECOND ADAPTIVE BEHAVIOR: ACTIVITY SELECTION

The success of SPT for habitat selection and the desire to address management questions related to within-day streamflow changes for hydropower generation prompted us to model the second adaptive behavior of salmonid fishes that we considered potentially important to population dynamics: selection of activity (feeding vs. hiding) in both day and night (Railsback et al. 2005). A number of field and laboratory studies have shown that trout select when as well as where they feed, as a trade-off between predation risk and energy intake: feeding at night generally reduces both growth and exposure to predation, while hiding results in no energy intake and very little exposure to predation. We modified inSTREAM to model habitat and activity selection by assuming trout select among four alternatives: feeding during the day and hiding at night, hiding during the day and feeding at night, feeding day and night, and hiding day and night. Model fish repeat this decision at the start of each diel (day, night) phase.

The fitness measure for this version of inSTREAM remains the expected probability of surviving predation and starvation over the next 90 days, but with separate estimates of survival for (1) day vs. night and (2) feeding vs. hiding. For example, expected survival of predation from current time t to the time horizon T, if the fish feeds during the day and hides at night, is modeled as $(S_{df}^{L} S_{nh}^{1-L})^{T-t}$, where S_{df} is the expected daily probability of surviving predation during daytime when feeding, S_{nh} is the expected daily probability of surviving predation when hiding at night, and L is day length (the fraction of a day that has daylight). At the start of each diel phase, individuals evaluate the fitness measure for each of the four activity combinations in each of the habitat cells available to them. They then select the activity and cell that provide the highest expected fitness.

Including the differences between day and night required modification of our representation of prediction. At the beginning of each diel phase, the model individuals still predict that current conditions will persist over the time horizon—but only for the diel phase currently starting. Because the current conditions do not represent the alternate phase, the individuals use memory of conditions during the previous phase as their prediction. For example, at the start of daytime a fish evaluates the fitness measure by assuming that the current daytime conditions will continue to occur each daytime over the time horizon and that conditions remembered from the previous night (the growth rate and predation survival probability, for both feeding and hiding, at the best cell available for each activity) will occur each night.

Adding activity selection to inSTREAM considerably increased its complexity, but it also increased the variety of realistic patterns of individual behavior and population dynamics that emerged from the model. These additional patterns included variation among individuals in diel activity, more fish hiding during the day than at night under certain conditions, less nocturnal feeding at higher temperatures, differences among life stages in nocturnal feeding, and the effects of food availability, competition, and physical habitat on when fish feed (Railsback et al. 2005). Inclusion of activity selection also appears important to the use of inSTREAM in environmental assessments: we have shown (Harvey et al. 2014) that populations in IBM simulations that included activity selection were less sensitive to reduced streamflow than populations with habitat selection capability alone. The ability to represent activity selection also has contributed to an understanding of basic ecological issues important to fisheries management, such as the role of food limitation in population dynamics (Railsback and Harvey 2011): the ability to trade off food for safety through activity selection allowed simulated trout to continue benefiting from increasing food availability at food levels much higher than most fish biologists would predict. Higher food availability allowed fish to feed less often and therefore expose themselves to less predation and live longer, resulting in higher abundance (figure 2.3).

Especially with activity selection, the use of SPT gives inSTREAM the ability to produce the trait-mediated indirect trophic interactions and nonconsumptive effects discussed in section 1.1 as characterizing how trade-off behavior affects ecology (Railsback and Harvey 2013). Trout IBM simulations have yielded (1) trait-mediated indirect effects of predation risk on food resources much stronger than density-mediated indirect effects, because fish increased nighttime feeding as predation risk increased; (2) strong bottom-up trait-mediated indirect effects, wherein increasing resource availability allowed fish to avoid predators by reducing feeding effort; (3) decreasing nonconsumptive effects of predators on trout as trout density increased, because competition for food reduced the opportunity for trout to avoid predators by reducing feeding effort; and (4) environmental effects on the strength of trait-mediated indirect interactions: indirect effects of

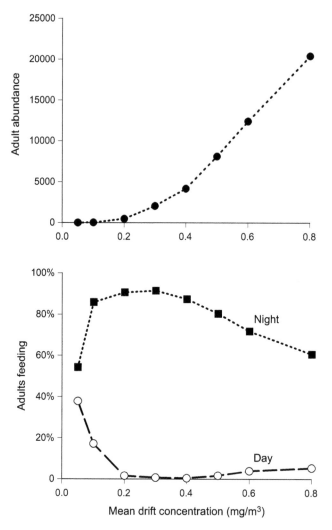

FIGURE 2.3. Results of a food limitation experiment using inSTREAM (redrawn results from Railsback and Harvey 2011). Top: As the concentration of food (invertebrates drifting in the water column, the primary food of adult trout) was increased tenfold over eight long-term simulations, the abundance of adult trout increased continuously. This result contradicts the traditional notion of food limitation, which posits that when food availability is sufficiently high it becomes unimportant and population size becomes "limited" instead by another factor such as predation. Bottom: Activity selection allows food to contribute to abundance even at high availability. The Y axis is the average percentage of adult trout feeding during any day or night. At low food availability, trout feed either during day or night. As food availability increases, trout rapidly shift to feeding almost exclusively at night, when both feeding success and predation risk are lower: the higher food concentration allows fish to obtain sufficient food while greatly increasing survival and, therefore, abundance. At even higher food availability, some trout further reduce risk by feeding only every other day.

trout on their food resource declined as temperature increased because trout had less scope to avoid predators by reducing feeding.

2.4 CONCLUSIONS

InSTREAM has evolved into a family of models, each focused on specific salmonid communities and management problems. These models have been applied at approximately 50 sites. Site-specific management applications have included assessment of alternative flow release schedules from hydropower dams, evaluation of habitat restoration programs (Railsback et al. 2013), analysis of small-stream diversions (Harvey et al. 2014), and design of habitat restoration projects. Chapter 7 discusses the most complex management application, which included the addition of a third adaptive behavior to address how stream habitat restoration would affect life history and populations of salmonids that decide adaptively whether to migrate to the ocean.

The models have also proved useful as virtual laboratories for exploring more general questions, of both management and theoretical importance, for which inclusion of adaptive trade-off behavior is probably critical. These questions have included how multiple stressors (physical habitat loss, increased turbidity, elevated temperatures) interact to affect populations (Harvey and Railsback 2007), how opposing effects of increased turbidity—reduced feeding success and reduced predation risk—interact to affect populations (Harvey and Railsback 2009; further discussed in section 10.4), how habitat fragmentation affects population persistence and size structure (Harvey and Railsback 2012), and (using an experiment that could not be conducted on real animals or in a model lacking adaptive individual behavior) how useful habitat selection models are for predicting population responses (Railsback et al. 2003).

InSTREAM's usefulness as a model of community ecology has also been demonstrated in several ways. The model can represent multiple, competing trout species that differ only in a few critical ways, such as reproducing at different times of the year. In a hypothetical ecotoxicology assessment, Forbes et al. (2019) used inSTREAM to predict the effects of a contaminant (an artificial estrogen compound) that reduces reproductive success of trout. The target species was an endangered cutthroat trout subspecies that co-occurs with introduced brown trout. Because of its detailed representation of life history and how individuals of the two species interact, inSTREAM could illustrate how the timing of contaminant loading, the seasonality of natural river flows (which dilute the contaminant), and each species' spawning time all interact to determine contaminant effects on the community. The assessment indicated that constant loadings had more severe effects on the native species, but that they could be offset by increased angler harvest of the introduced brown trout. The model also simply represents trophic levels both above and below trout. Railsback and Harvey (2013) illustrated the ability of inSTREAM to produce the complex, indirect trophic interactions that characterize systems of adaptive individuals (the tri-trophic model system introduced in section 1.1).

Despite our initial interest in management applications of the model, we quickly became aware of its potential to address the question of how adaptive behavior affects population and community ecology, which has been at the forefront of theoretical ecology. Our reading of the literature suggested that ecologists were making little progress modeling populations and communities of adaptive individuals. At the same time, we interacted often with individual-based modelers in many scientific disciplines and found that they too struggle with how to represent adaptive trade-off decisions in realistic systems with feedbacks and variable environments.

The motivation for this book, therefore, was the realization that the methods we developed for modeling habitat selection in trout essentially provided theory for adaptive trade-off decisions—SPT—that could be especially useful in IBMs with sufficient complexity to address real ecological and human systems.

CHAPTER 3

Introduction to State- and Prediction-Based Theory

3.1 WHAT IS SPT?

The fundamental concept of modern theory in behavioral ecology is that behavior acts to maximize a specific measure of fitness at a specific future time, and that this fitness measure incorporates multiple elements, such as the need to avoid predators, the need to avoid starvation, and the benefits of energy accumulation for reproduction. Maximizing a fitness measure forces an individual to make good trade-offs among its elements. When the fitness measure also considers the individual's current state, usually its energy reserves, the balance among fitness elements becomes state-dependent: hungry individuals take more risks to get food, while individuals with high reserves have more scope to avoid predation by feeding less often or in safer places.

This concept has been applied widely and successfully in DSVM, and we developed SPT as a way of using the same principle in IBMs when feedback from the behavior of other individuals, combined with unpredictable environmental conditions, make the assumption of optimality used by DSVM impossible.

To us, DSVM is such a powerful, general, and theoretically appealing approach that anyone attempting to model adaptive trade-off behaviors should become familiar with its literature (e.g., Mangel and Clark 1988; Houston and McNamara 1999; Clark and Mangel 2000) and with any applications of DSVM to the particular species and behaviors of interest to them. Many steps in using SPT are identical to those in DSVM, so much of the DSVM literature is directly applicable and we do not attempt to reproduce it here.

While SPT is based on DSVM, making it useful as across-level theory in IBMs requires that SPT differ from DSVM in three major ways. These differences change the decision-making and mathematical framework from one of optimization to one of approximation and updating, allowing individuals to continually adapt to changing and unpredictable conditions.

The first and most important difference between SPT and DSVM is updating instead of optimization over a specific time period. DSVM uses a one-time optimization to determine what an individual will do over an entire time period, including any changes in behavior it should make between the present and a future time horizon. Both DSVM and SPT use "time horizon" as the future date or time for which the fitness measure is evaluated. Because DSVM is individual-level theory, it can assume that the behavior of the individual being modeled does not affect future conditions, such as food availability and predator density. Further, DSVM assumes that future environmental conditions, including any effects of intraspecific competition, are known in advance. These assumptions are necessary to use a one-time mathematical optimization (often, dynamic programming) that "solves" what the individual's decisions should be over a future time period. In contrast, SPT cannot assume that behavior has no effect on future conditions and does not assume that future environmental conditions can be known in advance. Therefore, instead of one-time optimization, SPT uses *updating*: as time passes and conditions change, individuals reevaluate and change their behaviors. Updating allows individuals to adapt to changes in both their internal state and external conditions, whether those changes result from their own experience, the behavior of other individuals, or environmental variability. Instead of using DSVM to calculate what an individual should do over an entire future period, we use SPT in a simulation model to update decisions in each time step.

The second difference is that SPT explicitly represents how individuals predict conditions over future time. Usually, applications of DSVM implicitly assume that individuals just "know" the variables that represent conditions (e.g., food availability, predation risk) until the time horizon: they optimize their decision using perfect prediction. Perfect prediction is not feasible when future conditions are subject to feedbacks from behavior, and is certainly not a reasonable assumption for important conditions that depend on weather or other inherently variable processes. With SPT, prediction is represented explicitly, and imperfect prediction is normal. In fact, SPT allows us to investigate how important the accuracy of predictions is to successful adaptive behavior.

The third difference between SPT and DSVM is that SPT typically uses approximation in selecting the best choice of behaviors over time, instead of mathematical optimization. The most common approximation we use is that individuals make choices each time step assuming (incorrectly) they will use that option until a time horizon. Simplifications such as these can make SPT computationally tractable (and perhaps more realistic) for complex decisions that sometimes evaluate hundreds of alternatives.

3.2 FIVE STEPS FOR IMPLEMENTING SPT

To model populations of adaptive individuals, SPT is implemented using five steps, which we outline here. These steps include embedding SPT in an IBM that simulates the processes that drive behavior, both internal to the individual (e.g., feeding, energetics, fecundity) and external (e.g., food availability, competition, predation risks). Chapters 4 through 6 provide examples of SPT development, and chapter 8 provides guidance on each step and more detailed discussion of all the topics introduced here.

Step 1. Define the decision that SPT will model. Before we can model a decision, we must define it exactly: What entity makes the decision, in what context? What elements of fitness (survival of predation or starvation, reproduction) does the decision affect? This step is normally accomplished by designing the IBM that we use SPT within. Designing the population model to address specific problems in specific systems answers these questions that define the adaptive behavior.

Step 2. Design a fitness measure that represents how elements of individual fitness (size, survival, reproduction) at a future time horizon depend on individual state variables and how these elements combine into a single measure of future fitness. (The fitness measure corresponds to the "objective" concept in the ODD protocol commonly used to describe IBMs; Grimm et al. 2010.) This step closely parallels DSVM, so guidance on fitness measures from the DSVM literature (e.g., chapter 12 of Clark and Mangel 2000) is directly applicable.

Fitness measures are almost always probabilistic: they might represent an expected probability of surviving both predation and starvation until the time horizon, or an expected number of offspring at the time horizon that is equal to (a) the number of offspring that would be produced if the individual survived to reproduce, multiplied by (b) the probability of surviving until the time of reproduction. Therefore, a basic understanding of probability is essential for using SPT.

Designing a fitness measure includes specifying the time horizon. The DSVM literature and this book include time horizons such as the start of the next reproductive period, the assumed date of a life history transition, the estimated end of an organism's life span, and (as in the trout model of chapter 2) a "sliding" time horizon always n days away.

Step 3. Make explicit assumptions about how individuals predict future values of variables that affect the fitness measure. This step sets SPT apart from other behavior theory. Instead of implicitly assuming that individuals accurately know future conditions, we must model how they anticipate conditions as part of their decision-making process. Because this step is new, we give it more attention here and especially in section 8.4.

It is important to understand that this theory does not necessarily assume that organisms (or other kinds of individuals) consciously make predictions; it instead assumes that organisms have behaviors that can be modeled well by representing them as being based on explicit predictions. For simple organisms, we may be using our assumptions about prediction as a model of innate, reflexive behaviors.

At the other extreme, these assumptions may be a fairly close representation of how more conscious organisms actually do make decisions.

How to model prediction is a key question in the application of SPT. Several of the models we examine, starting with the trout model of chapter 2, illustrate that even very simple predictions that are always wrong can produce complex behaviors observed in nature. Yet recent studies of real organisms suggest that some individuals have capabilities similar to prediction that would improve simulated adaptive behavior if included in models (section 8.4.1).

This step also includes developing the submodels needed to calculate the fitness measure's value at the time horizon, using predicted future conditions, for each of the alternatives available to the individual. These submodels typically represent such processes as feeding or energy intake, how growth and reproduction depend on feeding and physiology, starvation mortality and how it varies with energy reserves, and how predation risk varies among alternatives.

The kinds of submodels addressed in this step are also required for DSVM, so the DSVM literature is instructive. However, DSVM and SPT also differ at this step. For example, DSVM does not rank each alternative by the value of its fitness measure, but instead uses dynamic programming to find one sequence of alternatives to use over time that maximizes the fitness measure. To avoid the need for dynamic programming optimization with SPT, we can instead use simplifying assumptions that, while also typically inaccurate, both produce good decisions and let us use more complex and realistic models of how internal and environmental variables affect future fitness. The assumption that individuals evaluate their fitness measure for an alternative by assuming that they will use only that alternative from the present until the time horizon is very useful because it simplifies evaluation of the fitness measure. This assumption is often wrong, but because we let individuals change their decision each time step as conditions change, it can still produce good decisions.

The benefit of giving up optimization in SPT is that it makes more complex and realistic submodels of fitness-related processes tractable. The need to solve its optimization presents a strong incentive to keep the fitness-related submodels of DSVM simple. One example, explored in section 4.5, is in modeling starvation survival: many DSVM applications use the very simplistic assumption that starvation happens if, and only if, energy reserves fall below a particular threshold. With SPT, we can easily assume that the risk of starvation increases gradually as energy reserves decrease, and this difference strongly affects simulated population dynamics. This is one illustration of how using approximation instead of optimization in SPT can allow us to make models that may better represent real decisions.

Step 4. Develop a decision algorithm to select a good alternative. Now that we have a way to evaluate the fitness measure at the time horizon for each alternative, we design how the individual decides which alternative it will actually implement. Ideally, information on the capabilities of the individuals making the decision can be considered in choosing a decision algorithm. When modeling natural behavior, it is often reasonable to assume individuals can identify and use the best alternative; but when modeling responses to highly unpredictable or novel events (e.g., responses to contemporary human activities), a more approximate decision algorithm may be appropriate.

The decision algorithm we use throughout this book is simply to evaluate the fitness measure for each alternative and choose the alternative with the highest value. This approach assumes that individuals have the time and ability to evaluate each alternative (meaning that we need to be careful in defining how many alternatives they consider) and identify the best one. This approach is so natural that many modelers use it without even thinking of it as a decision algorithm, and it corresponds with the understanding of how animals actually make decisions described in section 1.4. The decision theory literature, however, includes a variety of models of how individuals select among alternatives when they have a measure (like our fitness measure) of the benefits provided by each. Much of the decision theory literature addresses "efficient" methods that reduce the cost (e.g., in information-gathering) of making decisions while still producing satisfactory outcomes.

Step 5. Implement the previous steps in an IBM, which simulates the survival, growth, reproduction, etc. that emerge as individuals make and routinely update their decisions over time. The higher-level dynamics we are interested in as population and community ecologists then emerge from the fates of the IBM's individuals. SPT is theory for how adaptive behavior affects population dynamics, so while it could be used just to model individual behavior, its power is in representing behavior in simulation models of interacting individuals.

Implementing SPT in an IBM includes tasks such as developing and testing model software, calibrating and testing the model, and analyzing its sensitivities and uncertainties. In this book we do not focus on aspects of individual-based modeling other than those related to behavior. Among the extensive literature on IBMs, Grimm and Railsback (2005) provide broad coverage while Railsback and Grimm (2019) provide more hands-on guidance on model design, implementation, and analysis.

3.3 A LOOK AHEAD

This chapter is intended to provide a quick overview of SPT and its use in IBMs, as background for the example applications we provide next. Chapter 8 collects our experience to provide much more detail and guidance for each of the five steps.

Our discussion here about how SPT depends on inaccurate predictions and approximations no doubt raises the question of how successful its adaptive trade-off decisions could possibly be. The trout model examined in chapter 2 is among the most detailed and mechanistic ecological IBMs developed so far, so it illustrates that SPT is not only practical in complex models (with sometimes many thousands of individuals making decisions by evaluating sometimes hundreds of alternatives) but also capable of producing the behavior-driven trophic interactions defined in section 1.1. However, for the next three chapters we use simpler models to keep the focus on the adaptive behavior theory. Chapters 4, 5, and 6 each look at models simple enough that we can compare decisions produced by SPT to those produced by DSVM with dynamic programming optimization.

The most important way that we address how successful SPT can be at modeling adaptive behavior comes in chapter 9. There we describe and illustrate a hypothesis-testing cycle we can use to test, improve, and document the success of adaptive behavior theory in individual-based population models.

CHAPTER 4

A First Example:
Forager Patch Selection

4.1 OBJECTIVES

Here we start illustrating how to use SPT in detail, by applying it to an extremely simple model of how hypothetical foragers select among patches that differ in both food availability and predation risk. A second important objective of this chapter is to evaluate the quality of decisions made via SPT. We address this question by comparing SPT's results to those produced by DSVM: the SPT patch selection model presented here is based on the DSVM patch model of Mangel and Clark (1986).

Our third objective is to start showing how SPT allows us to make models less simple and more realistic and capable. We do this by modifying the model in two ways that would make it far more difficult to solve via optimization. First, we change the model from an individual-level model into an IBM of interacting individuals that contains feedbacks of behavior, via competition for food. Second, we make starvation mortality a gradual function of energy reserves instead of a binary function. For each of these two new versions of the model, we examine the new system dynamics that emerge from individual behavior. In particular, we look at the patch selection IBM as a tri-trophic system and examine the extent to which it produces the behavior-driven trophic interactions identified in section 1.1.

This book's web site offers software implementations of all versions of the patch selection model (see the preface for information).

4.2 THE MODEL

Mangel and Clark (1986) used patch selection as their first example of how to model adaptive trade-off decisions using DSVM. We reinterpret their model as an IBM, making no changes to its assumptions except for how the trade-off decision is modeled.

For this and all the example models we present, we use the ODD (Overview, Design concepts, and Details) protocol for describing individual-based models (Grimm et al. 2010; Railsback and Grimm 2019). The ODD protocol provides a standard organization for model descriptions, and its "design concepts" element provides a framework for thinking about (and designing as well as describing) aspects of IBMs that are not well captured by mathematics. ODD is now widely used in ecology and in fact has been a major factor unifying the use of IBMs in ecology with similar approaches in other fields (Vincenot 2018). ODD has seven elements:

1. **Purpose and patterns**. The purpose of this model is simply to explore ways of modeling foraging decisions and their effects on individual fitness, represented via survival probability and energy reserves. Because this model does not represent a specific species or system, the patterns we use to evaluate it can only be general expectations of how foragers behave under predation risk: that they select habitat to maintain a balance between food intake and predation risk, and that this balance depends on the state of their energy reserves.

2. **Entities, state variables, and scales**. The model includes two kinds of entities: patches and foragers. There are five patches, each with only two variables: the daily probability λ that a forager on the patch will find a food item, and the daily probability P of a forager surviving predation there. The five patches present a range of food–risk trade-offs. In order, patches 1–5 have λ values of 0, 0.20, 0.35, 0.35, and 0.50 and P of 1.0, 0.95, 0.85, 0.75, and 0.60. Therefore, patch 1 is a refuge providing no food intake but no risk of predation; patch 5 provides the highest probability of feeding success but the lowest probability of survival. Patch 4 is notable because it provides the same feeding success probability as patch 3 but lower survival.

   The model's foragers represent animals that select foraging habitat by making trade-off decisions. Only one short life stage is simulated; it can be thought of as an intermediate life stage of fixed duration that an animal must survive before advancing to adulthood. Foragers have only one state variable, their energy reserve χ, which is in arbitrary units.

   The model simulates a life stage of duration T days, normally 10. Each time step represents one day, so the simulated time t increments from 1 to 10 days.

3. **Process overview and scheduling**. On each day, foragers execute the following five actions. Each forager executes all five actions before the next forager executes any action. The order in which foragers execute is randomized each time step.

a. *Habitat selection* is the choice of which of the five patches to forage in. Habitat selection uses SPT as explained in element 7 (Submodels).

b. *Feeding* is modeled as a stochastic event: with the probability of success equal to the value of λ in its chosen patch, the forager does or does not succeed in obtaining food.

c. *Growth* is a simple energy balance. If the forager successfully feeds, its value of χ increases by the food value Y (2 energy units). Regardless of feeding success, χ is then decreased by a maintenance cost α (1 energy unit).

d. *Predation survival* is also a stochastic event: the forager survives with probability equal to the value of P at its selected patch.

e. *Starvation survival* is a very simple mechanistic process. We first follow Mangel and Clark (1986) by treating starvation survival as a binary function of energy reserves: an individual starves to death if χ declines to zero. However, in section 4.5 we also explore the alternative assumption that starvation probability increases continuously as energy reserves decrease.

4. **Design concepts**. This ODD element uses 11 "design concepts" to describe essential characteristics of the IBM not captured well by equations or algorithms.

a. *Basic principles*: The model implements the basic foraging theory principle that organisms have behaviors that promote individual fitness, where fitness can be represented by measures of future reproductive potential, such as survival to a time horizon, so behavior can be modeled by assuming individuals act to increase or maximize a mathematical measure of future fitness (e.g., Mangel and Clark 1986; Brown et al. 1999).

b. *Emergence*: The main model result that emerges from adaptive individual behavior is survival: the fraction of foragers that succeed in avoiding predation and starvation until the end of the life stage. Survival depends on an important secondary result, the pattern of patch selection over time: how many individuals use which patches on each time step.

c. *Adaptation*: Model individuals have one adaptive behavior: their decision of which patch to occupy in each time step. They make this decision by choosing the patch that provides the highest value of a fitness measure.

d. *Objectives*: The specific objective (in this case, "fitness measure") used to evaluate the alternative patches is an approximated probability of surviving predation and starvation until the time horizon. It is explained in detail under Submodels below.

  e. *Learning*: No learning is modeled.

  f. *Prediction*: The foragers, in selecting a patch, use the explicit predic-
     tion that patch characteristics λ and P will remain unchanged until
     the time horizon. This prediction is accurate in the first version of the
     model. However, it is not accurate for subsequent versions that assume
     individuals deplete and compete for food (or in models with environ-
     mental variability).

  g. *Sensing*: Foragers accurately sense the values of λ and P in all patches.
     They also can accurately sense their internal energy reserve χ.

  h. *Interaction*: The first version of the model assumes no interaction
     among the foragers: they do not compete or otherwise affect each other.

  i. *Stochasticity*: Feeding success and predation survival are modeled
     as stochastic processes, as a way of making these processes variable
     among individuals without representing why they vary.

  j. *Collectives*: There are no collectives (aggregations of individuals that
     affect individual behavior and population dynamics).

  k. *Observation*: The distribution of foragers among patches and over time
     can be observed visually via a graphical interface and from a file that
     reports the location and energy reserve of each forager on each time
     step. Mangel and Clark (1986) reported patch selection results as the
     calculated probability of an individual surviving to the time horizon.
     For comparison to those results, we simulate 10,000 individuals and
     report the fraction surviving.

5. **Initialization**. The model is normally initialized with 100 foragers, all
   with the same initial energy reserve χ_0, which is normally 5 units. Foragers
   are not given an initial patch; the habitat selection action on the first time
   step determines their location at $t = 1$.

6. **Input data**. The model uses no time-series input.

7. **Submodels**. The only submodel not fully described above is habitat selection. Mangel and Clark (1986) modeled habitat selection as a one-time
   optimization that determines which patch the forager should use on each
   of the T days, by maximizing a fitness measure representing the probability of surviving predation and starvation over the entire life stage. The
   probability of surviving predation is simply the product of the values of P
   for all the patches occupied during the 10-day time horizon. The probability of surviving starvation is simply 0.0 if energy reserves reach zero and
   1.0 if they do not.

   Our SPT approach instead lets each forager update its patch decision
   each day, with the decision depending on its current energy reserve. They
   do so by selecting the patch that provides the highest value of a fitness

measure F equal to the expected probability of surviving predation and starvation until the time horizon (T, the final day), using one simplifying assumption and the prediction stated above. The simplifying assumption is that the forager would remain in the selected patch for the $T - t$ days remaining in the life stage. This assumption clearly is often wrong, as the foragers can move among patches each day, but without it we would need to use dynamic programming as Mangel and Clark (1986) did.

With this time horizon, simplification, and prediction, the fitness measure F is the probability of surviving until time T if the forager stays in the patch until that time horizon and if the daily feeding success probability λ and predation survival probability P remain constant. This expected survival probability has terms for both predation and starvation. Expected survival of predation is simply $P^{(T-t)}$. Expected survival of starvation is also simple because starvation is binary: we can determine whether or not the expected energy reserve at the time horizon is positive and set starvation survival to 1.0 or 0.0 accordingly. Under our assumptions, the expected value of χ at time T is equal to $\chi_t + (T - t)(\lambda Y - \alpha)$. Therefore, if $\chi_t + (T - t)(\lambda Y - \alpha) > 0$, then $F = P^{(T-t)}$; if not, then $F = 0$.

4.3 RESULTS AND COMPARISON OF SPT TO DYNAMIC STATE VARIABLE MODELING

Now that we have turned the DSVM patch selection model of Mangel and Clark (1986) into an IBM with individuals using SPT to select patches each day, let us compare the results of the two models. We can compare both the patterns of habitat selection over time and the survival rates that emerge from the patch selection decisions.

Results of this simple example illustrate both similarities and differences between DSVM and SPT. As expected, and like the DSVM results, no SPT foragers ever used patch 4, which has the same food availability as patch 3 but higher predation risk. For 10-day life stages, Mangel and Clark (1986, table 3) indicate that patch 3 is optimal for early time steps; similarly, as shown in figure 4.1, SPT foragers always start in patch 3, and most stay there until they can move to the lower-risk patches late in the simulation. Also as in the DSVM results, all SPT foragers surviving the simulation ended with the minimum of 1 unit of energy reserves.

However, SPT foragers exhibited variable behavior as a consequence of their luck in finding food each day (figure 4.1). Those who failed to find food had lower reserves and therefore assumed higher predation risk to catch up (e.g., the

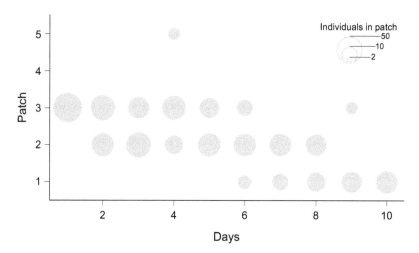

FIGURE 4.1. Patch selection in the forager model over a 10-day simulation. Symbol area is scaled to the number of foragers, out of an initial 100, occupying the patch indicated on the Y axis.

occasional use of patch 5). On the other hand, the lucky foragers who consistently found food were often able to retreat to patch 2 or even to the patch 1 refuge.

As with DSVM, the SPT foragers' decisions varied with the duration of the life stage and initial energy reserves. When we simulated a 20-day life stage with the standard χ_0 of 5 units (figure 4.2), most foragers used patch 5 early, because the probability of starving over a long time horizon is high without relatively high reserves. Initializing simulations with χ_0 of 2 and 8 units produced the expected differences in patch selection (figure 4.3). Low initial energy reserves caused foragers to use the high-risk patch 5 early in the simulation (and far fewer of them to survive), while high initial reserves allowed foragers to start in patch 2 and remain in the lower-risk patches. Mangel and Clark's DSVM solution was for foragers to start in patch 3 when χ_0 was 2, and (like our results) to start in patch 2 when χ_0 was 8.

Survival results of our SPT simulation resemble those of Mangel and Clark's (1986) DSVM optimization (figure 4.4), over 10 simulations varying χ_0 from 1 to 10 energy units. Mangel and Clark report the probability of survival for an individual that optimally selects patches, whereas we report survival rates observed in simulations with 10,000 initial foragers that experience stochastic feeding success and hence vary in patch selection and predation risk. SPT produced survival rates that were, as expected, always less than the survival probabilities in the DSVM optimal solution but only an average of 13% lower in the 10-day simulations. The difference between SPT and DSVM was much greater when χ_0 was low and, therefore, the survival problem was much harder; when χ_0 ranged from 1

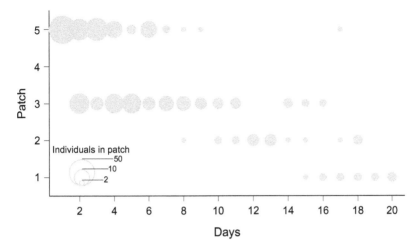

FIGURE 4.2. Patch selection results for a 20-day simulation; format as in figure 4.1.

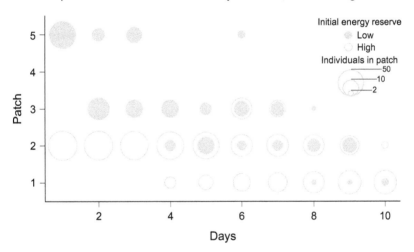

FIGURE 4.3. Patch selection results with initial energy reserves χ_0 lower (2 units) and higher (8 units) than the standard value of 5.

to 4, survival was less than 14% even with the DSVM optimization. Under these conditions, the approximations used by SPT appear to have substantial (though possibly realistic) impacts on survival. On the other hand, when χ_0 was between 5 and 10 units, survival using SPT averaged only 5% lower than with DSVM.

In the 20-day simulations, SPT also produced results similar to DSVM (figure 4.4). SPT produced survival probabilities 39% lower than DSVM, while average survival dropped to only 11% of 10-day survival. The comparison of 10-day and

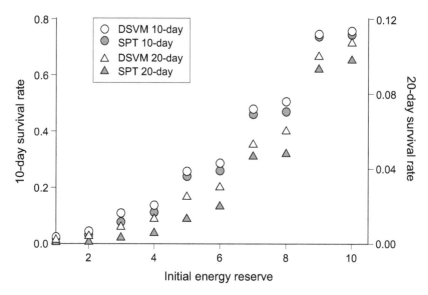

FIGURE 4.4. Comparison of survival in SPT simulations to optimal survival reported by Mangel and Clark (1986) for a DSVM version of the patch selection model.

20-day results again shows the difference between SPT and DSVM increasing as survival becomes more challenging. As in the 10-day simulations, the difference between DSVM and SPT survival was lower when χ_0 was higher. For simulations with χ_0 of 7–10 units, SPT survival averaged only 12% less than with DSVM.

These results indicate that SPT, with its approximations, produces behavior qualitatively similar to the behavior produced by DSVM optimization, while letting individuals realistically adapt their patch selection decisions to their individual experience with feeding success. SPT also produced survival rates close to the optimal rates except when survival was very difficult, e.g., when fewer than 10% of individuals survived the short life stage.

But we have not yet addressed the real promise of SPT: to examine the consequences of individual trade-off decisions at the population level and to predict and explain complex ecological dynamics. For example, does this simple forager patch selection model produce the top-down and bottom-up trait-mediated indirect interactions (TMIIs) characteristic of the tri-trophic model system we introduced in section 1.1? The increase in forager survival with increasing χ_0 is promising: this response hints that the model can produce bottom-up TMIIs. Higher food availability, like higher χ_0, should allow foragers to adapt their patch selection to reduce predation rates.

We can easily test whether the patch selection model produces TMIIs by conducting additional simulation experiments. First, we tested for bottom-up TMIIs

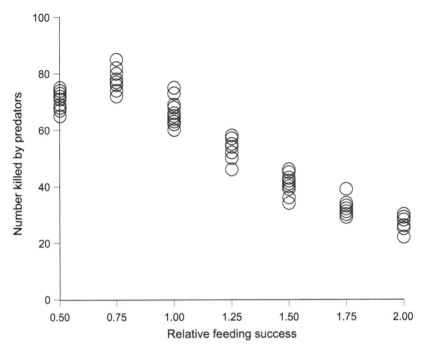

FIGURE 4.5. Response of the number of foragers killed by predators to food availability. Each symbol represents one of ten replicates for each food availability scenario. X values represent food availability as multipliers of standard feeding success probability.

by varying food availability and examining the response in predation on foragers. In seven food availability scenarios, we varied the feeding success probability λ in all patches from 50% to 200% of its standard value, in increments of 25%. For each of these scenarios, we counted the foragers killed by predators. The results, shown in figure 4.5, indicate that overall the model indeed produced this complex dynamic: the number of foragers killed by predators declined steeply as feeding success increased. However, when feeding success increased from 50% to 75% of standard values, predation actually increased because far fewer foragers died of starvation, leaving more for predators to consume. As food availability increased further, however, the number of foragers continued to increase but the number eaten by predators decreased sharply because of their adaptive behavior: with higher food availability, foragers moved to safer patches where they could still meet their energy needs.

Next we tested for top-down TMIIs by varying predation risk and examining food consumption. We used seven predation risk scenarios that varied the daily probability of predation mortality $(1 - P)$ in all patches from 50% to 200% of standard values, and counted the total number of food items consumed by foragers. As

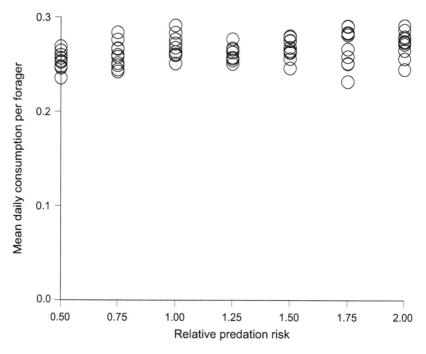

FIGURE 4.6. Response of food consumption (items per forager per day) to predation risk, with ten replicates of seven risk scenarios. X values represent predation risk as multipliers of standard mortality probability values.

predation risk increases, total food consumption will no doubt decrease because fewer foragers will be left alive to eat. Instead, we looked at per capita consumption: How many food items per day did each surviving forager catch?

In contrast to bottom-up TMII, the model produces no top-down response of food consumption to predation risk (figure 4.6). In fact, the number of foragers alive at the end of the life stage decreases linearly with predation risk, confirming that the foragers cannot adapt their patch selection behavior to mitigate increases in predation risk. Why? We explore that question in section 4.5.

4.4 VERSION 2: FORAGING WITH COMPETITION

The patch selection model described and analyzed in sections 4.2–4.3 is hardly an IBM: IBMs are population (or community) models, and this model has so far only simulated individuals that do not interact. The "population" dynamics were driven entirely by external factors: food availability and predation risk. Now we will modify the model to include one of the primary internal factors governing many populations: competition.

Adding competition is a critical step because it introduces feedbacks. When individuals compete for food, the alternatives available to each individual and its behavior now depend on what the other individuals do. These kinds of feedbacks make DSVM and game theory impractical for population modeling because feedbacks make the number of potential model states extremely large and optimization infeasible. In contrast, SPT can typically handle competition and its effects on behavior and emergent population dynamics with literally no change to the behavior model.

To illustrate the ability of SPT to function when behavior is subject to feedbacks, we modified the patch selection model to include competition for food. This simple modification results in the following changes to the ODD model description of section 4.2:

2. **Entities, state variables, and scales**. The patch variable λ, the probability that a forager will find a food item, is no longer a static variable but changes within each time step.

3. **Process overview and scheduling**. (a) The *feeding* action is modified to represent food consumption. Whenever a forager succeeds in finding food, the value of λ for its patch declines by 0.01, making feeding success less likely for subsequent foragers on the same day. (b) A new *food regeneration* action is added to the schedule as the first action in each time step. This action resets the value of λ to those used in the first version of the model: 0, 0.20, 0.35, 0.35, and 0.50 for patches 1–5.

4. **Design concepts**.

   Emergence: The population survival rate now emerges from the initial population density as well as from adaptive patch selection behavior. Population density affects competition, which affects patch selection, which determines survival.

   Prediction: No change is made to how foragers predict future feeding and predation probabilities: they still assume that the probabilities occurring at the current time step will persist until the time horizon. But this prediction is no longer accurate for feeding success, because the probability of finding food depends on how many other foragers have selected a patch.

   Interaction: Foragers now interact with each other indirectly via competition: feeding success by one individual reduces the probability of success for subsequent individuals. The scheduling assumption that foragers execute their actions in randomized order each time step simulates

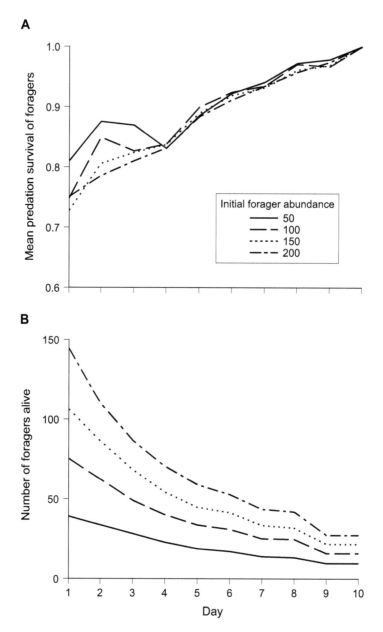

FIGURE 4.7. Results for the patch selection model with competition, over a 10-day simulation. Lines depict the means over 10 replicate simulations. (A) Mean predation survival P of patches selected by foragers. P varies from 0.6 in patch 5 to 1.0 in patch 1. (B) Forager abundance. With 200 initial foragers, > 50 foragers die on day 1.

"scramble" competition: all foragers have equal chances of feeding earlier vs. later each time step.

With these modifications, we executed 10 replicates each of simulations with 50, 100, 150, and 200 initial foragers. The introduction of competition produces believable feedbacks in the behavior produced by SPT. Higher initial forager density causes foragers to spread out among more patches, usually those with higher food and higher risk (figure 4.7A), resulting in lower survival (figure 4.7B). Early in the simulations, competition for food in patch 3 causes foragers to often use patch 4 and even, at high initial abundance, patch 5. However, by day 4 the higher mortality in the high initial abundance scenarios eliminates the differences among scenarios in patch selection.

Is the behavior of switching to riskier patches as food is depleted in the best patch, even at the start of a simulation when energy reserves are high, realistic or reasonable compared to using a safer patch and hoping for feeding success later? Why does it happen? These questions relate to how we model starvation, which we address next.

4.5 VERSION 3: CONTINUOUS STARVATION RISK

In section 4.3 we saw that the patch selection model does not produce characteristic top-down TMIIs: per-forager food consumption did not respond as predation risk varied. In section 4.4 we saw that competition causes foragers to switch to riskier patches with more food instead of to safer patches, even when their energy reserves are relatively high. Both of these questionable outcomes appear to result from the model's treatment of starvation as a simple binary process: individuals starve immediately if their energy reserve falls to zero, but otherwise they have no risk associated with energy level.

This simple representation of starvation appears to be common in DSVM models, presumably because it makes the dynamic programming optimization easier. However, this assumption is clearly not realistic: instead of suddenly dying, real organisms gradually become more susceptible to starvation and disease as their energy reserves decline (we discuss this further in section 8.4.3). More important than being unrealistic, the assumption allows model individuals to make only crude trade-offs between predation risk and food intake.

It turns out that one major advantage of SPT is that it easily accommodates more sophisticated representations of starvation and other negative effects of low energy reserves. Because SPT does not require dynamic programming, more complex models of starvation are mathematically and computationally tractable.

We now illustrate this advantage of SPT by modifying the patch selection model to make starvation risk a continuous instead of binary function of energy reserves. Like the second version of the model, this version includes competition. Two additional changes to the ODD model description define how we now model starvation and its effects on behavior.

3. **Process overview and scheduling**. In the *starvation survival* action, survival is now represented as a Bernoulli trial: whether a forager survives starvation each day is a random event with probability S of surviving. S is a function of the individual's energy reserve χ: $S = 1.0$ when χ is greater than or equal to 1, and $S = \exp(\frac{\chi-1}{4})$ when χ is less than 1. Therefore, individuals do not automatically die when χ is zero, but their probability of survival decreases as reserves fall below 1 ($S = 0.78$ when χ is 0, $S = 0.61$ when χ is -1, etc.).

7. **Submodels**. The habitat selection submodel is modified so that the fitness measure F reflects the change in the formulation of starvation survival. F is now the product of expected survival of starvation and expected survival of predation until the time horizon. Expected survival of predation remains $P^{(T-t)}$.

   Expected survival of starvation until T (S_e) is determined by calculating, for each remaining day i, the expected energy reserve (χ_i) and the resulting daily probability of survival S_i. The values of χ_i are calculated by assuming energy intake each day is λY, using our prediction that future probability of feeding success remains the same as the current value of λ. The value of S_i is calculated from χ_i using the function defined above in the starvation survival action. The values of S_i for each remaining day are then multiplied together. This assumption is expressed mathematically as

$$S_e = \prod_{i=t}^{i=T} S(\chi_t + ([T-t][\lambda Y - \alpha]))$$

where S is the function relating daily starvation survival to energy reserve.

This version of the model does produce top-down indirect effects on forager survival and food consumption. We repeated the experiment that produced figure 4.6, varying predation risk in all patches from 50% to 200% of standard values, without competition. Because foragers now have greater ability to balance predation and starvation, per capita food consumption decreases as predation risk increases, as expected (figure 4.8). Because starvation becomes a very high risk as χ becomes more negative, the ability of foragers to compensate for predation risk by using less productive patches decreases at high risk levels.

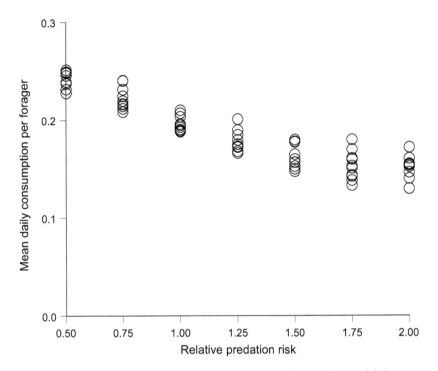

FIGURE 4.8. Response of food consumption to predation risk, with starvation modeled as a continuous function of energy reserves. Compare to figure 4.6.

Making starvation continuous instead of binary also produces more intuitive results when we include competition. Now, on the first few days, foragers never use the highest-risk patches except at very high initial densities; instead, as food availability declines they remain in lower-risk patches and sometimes even use the patch 1 refuge (figure 4.9A). Because starvation is not guaranteed if χ falls to zero, foragers are not forced to take extreme risks before their energy reserves are depleted. Instead of initially high predation mortality, the improved ability to trade off predation and starvation causes the forager population to experience steady mortality due to both causes (figure 4.9B).

4.6 CONCLUSIONS

Our main objectives in this chapter were to illustrate the application of SPT to a specific model, and to compare it as directly as possible to DSVM. We made the comparison by turning the patch selection problem that Mangel and Clark (1986)

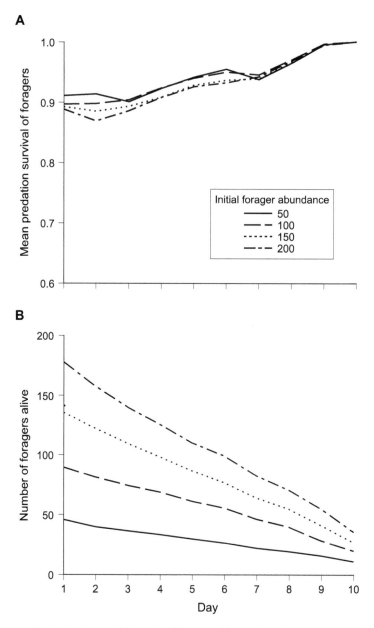

FIGURE 4.9. Patch selection model results with competition and continuous starvation risk, for comparison to figure 4.7.

used as an early example of DSVM into an individual-based population model in which individuals use SPT to select patches.

This first simple example uses two common techniques of SPT. First, individuals predict conditions using the simplest approach that lets them adapt to changing conditions: by simply assuming that current conditions will persist until the time horizon. This prediction can be completely accurate for variables like P that do not vary over time. But for variables subject to feedbacks of individual behavior—such as λ when we included competition for food—or time-varying environmental variables, this prediction can be wrong and yet useful for modeling behavior.

The second typical SPT technique we illustrated is to simplify evaluation of the fitness measure for a particular decision alternative (here, a patch that might be selected) by assuming that the individual will use the alternative (occupy the patch) until the time horizon. This assumption is also often wrong, yet useful.

Theory based consciously on assumptions that we know are often wrong inevitably raises the question of how good the decisions produced by the theory can be. We were able to address this question by comparing results of SPT to the DSVM optimal solution for the simplest version of the patch selection model. We found that SPT produces close approximations of the optimal behavior, although the differences between SPT and DSVM become greater, not surprisingly, in conditions where survival is very difficult.

The payoff for accepting the nonoptimality of SPT is that it lets us easily include more of nature's complexity in our models. We easily modified the patch selection IBM to include competition and then a more realistic representation of starvation survival, with little additional complexity or computational effort. Both of these changes caused the model to produce patterns, such as top-down TMIIs and use of safer instead of riskier habitat when energy reserves are still high, that intuitively seem more realistic. So while the SPT model achieved reasonable but suboptimal solutions to a very simple problem, it produced successful adaptive behavior in less simplistic, more realistic situations.

Our objectives for this chapter required the use of an extremely simple model system. In the next chapter, we look further into the benefits and challenges of SPT by modeling a more complex and realistic problem—one also previously addressed with DSVM.

CHAPTER 5

A Second Example:
Vertical Migration and
Reproductive Effort in *Daphnia*

5.1 OBJECTIVES

In this chapter we use SPT to model specific behaviors of a real species that have been both observed empirically and modeled previously using DSVM. Modeling a specific system, unlike the hypothetical foragers of chapter 4, means that we can compare our results to a variety of patterns observed in real organisms as well as in the DSVM optimization.

In this example, we examine *contingent* adaptive behaviors—two decisions that strongly affect each other, in this case via complex feedbacks. Specifically, we model a habitat selection decision and a life history decision: where to forage and how much assimilated mass to allocate to reproduction instead of growth. Habitat selection determines resource assimilation and survival probability, which affects the relative fitness of alternative resource allocation strategies, while the allocation of resources to growth vs. reproduction determines future size, which strongly affects habitat selection.

The habitat selection behavior we model is vertical migration (VM), a common and widely studied behavior of some aquatic animals. Animals from zooplankton to large predatory fish such as cod migrate up and down in the water column in daily cycles. While VM can be influenced by metabolic benefits linked to vertical temperature gradients (e.g., Wurtsbaugh and Neverman 1988), in many instances it is assumed to be an adaptive behavior trading off food intake and predation risk. This trade-off is driven by light, similar to adaptive trade-offs by terrestrial animals that determine how much time to spend feeding during the day versus at night. During the day, light intensity is high near the water surface but decreases with depth, while of course there is much less light at night. Light makes it easier for visual feeders to find food; such organisms can expect higher food intake, and their prey can expect greater risk, high in the water column during the day. For organisms like zooplankton that are both visual feeders and prey of visual feeders, being high in the water column during the day provides high intake and high risk, while being low during the day provides low risk but low intake. At night, depth has little effect on intake and risk, because it is darker everywhere (but still sometimes light enough for feeding at a reduced success rate near the surface). Daily VM therefore can reduce risk while allowing animals to obtain sufficient food.

The trade-offs driving an organism's VM depend on the characteristics and behaviors of its food and predators. Here we model a zooplankton species that consumes primary producers having little ability to migrate. Therefore, we can assume the zooplankton's food availability is highest near the surface around the clock. Species with mobile prey that also migrate have a more difficult problem because prey availability can change over time as well as with depth. Likewise, the predation risk that a migrating organism is trying to avoid may or may not vary with time or depth. The organism we model here is at risk from both invertebrate predators that are not visual feeders and visual-feeding fish that are a higher risk during the day and nearer the surface.

VM has long been recognized as an easily observed, interesting, and economically important trade-off behavior: it has been widely studied and modeled (e.g., McLaren 1963, Enright 1977, Gliwicz 1986). The Theoretical Ecology Group at the University of Bergen, Norway, has a particularly long history of modeling VM using DSVM and other methods (especially the artificial evolution of behavior discussed in section 1.3). Here we base an SPT model of VM on a DSVM model developed at Bergen by Øyvind Fiksen (1997). The model organism is *Daphnia magna* (hereafter *Daphnia*), a common zooplankter.

Fiksen's model includes a contingent behavior linked to VM: the decision of how to allocate assimilated resources to either growth or reproductive output. *Daphnia*, like many animals exhibiting indeterminate growth, appear to make this reproductive allocation decision as an adaptive trade-off. Allocating more resources to growth in size allows an individual to reach large size sooner, and large size reduces predation risk and increases reproductive output. Allocating resources to earlier reproduction reduces growth but produces offspring earlier, making it less likely that the individual dies without reproducing. In the unrealistic situation we simulate of a population growing without competition, earlier reproduction also conveys the fitness advantage of giving the offspring more time to produce additional generations. Clearly, the allocation producing highest fitness depends on both the rate of resource assimilation and predation risk— including how risk varies with size. Consequently, the best reproductive allocation can depend on the VM behavior and the best VM behavior can depend on the reproductive allocation.

Our objectives in this chapter are to show how important real behaviors like VM and reproductive allocation can be modeled via SPT, and to test how well SPT-based models compare to Fiksen's DSVM model in their ability to reproduce key patterns in these behaviors. We focus just on the second and third steps for using SPT defined in chapter 2, developing three versions of SPT with increasingly complex fitness measures and predictions. We then compare how well these versions reproduce patterns observed in real *Daphnia* or produced by the DSVM model. While our SPT models still use simple predictions and approximations of future fitness, some of them are more readily implemented using an algorithmic approach instead of mathematical equations.

We start by describing the model of individual *Daphnia* and their environment, including the observed patterns that the model is designed to explain. Then we present a series of ways that the two contingent adaptive behaviors can be modeled using SPT, evaluating each version by how well it can reproduce both observed patterns and patterns produced by the DSVM model. These analyses let us draw conclusions about how different levels of approximation in SPT affect its ability to produce realistic or useful adaptive behavior in an IBM.

Readers are reminded that they can read and execute computer code for all the model versions in the chapter at this book's web site.

5.2 THE MODEL

This model is based directly on the zooplankton model of Fiksen (1997), although below we note several differences. Also, Fiksen's model represents the behavior of an individual organism, so it does not represent the population-level processes that we introduce. In this section we describe our model in the ODD format, except for the adaptive behavior submodel. Three subsequent sections each describe and explore an alternative application of SPT to the adaptive behaviors.

1. **Purpose and patterns**. The purpose of this model is to explain and understand diurnal vertical migration (VM) in zooplankton and how it interacts with a life history trade-off, the allocation of mass to reproduction or growth. The model is based explicitly on the cladoceran *Daphnia magna*.

    We evaluate our model by its ability to reproduce three patterns. The first two are of primary importance because they were observed in extensive laboratory experiments by Loose and Dawidowicz (1994) and appear driven directly by the risk-growth trade-off (figure 5.1). We have less expectation that the third pattern will be reproduced because it reflects a secondary effect of the trade-off (via the effects of VM on energy intake)

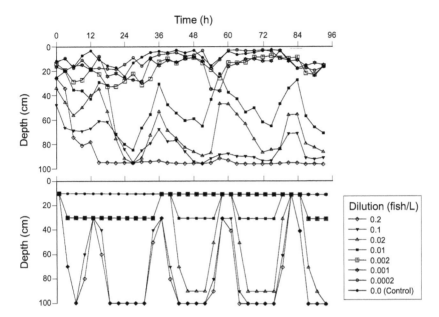

FIGURE 5.1. Vertical migration as a function of fish concentration, in (top) laboratory experiments of Loose and Dawidowicz (1994) and (bottom) the DSVM model of Fiksen (1997). Reprinted with permission from Fiksen (1997). The X axis is time in hours after experiments started at midday. Data lines are reproduced exactly as they appear in the source material. Axes and text labels have been redrawn for legibility.

and because it was not examined in the laboratory study but only observed in Fiksen's DSVM.

Pattern 1: Response of VM to predation risk. This pattern reflects how *Daphnia* VM in the laboratory changed as the perceived risk of predation by fish increased. In the laboratory experiments, perceived risk was increased by adding fish kairomones—fish-produced chemicals that *Daphnia* can use to sense fish density—and perceived risk was quantified as fish concentration (fish/L). With no perceived risk, *Daphnia* stayed near the surface throughout the diurnal cycle. At low fish concentrations, *Daphnia* remained near the surface or migrated only to shallow depths. With high risk, *Daphnia* stayed near the bottom. But at intermediate risk, *Daphnia* exhibited VM, with their mean elevation in the water column low during the day and rising toward the surface at night. And at the lowest fish concentration producing VM, *Daphnia* did not begin migration until growing to a threshold body size.

Pattern 2. Selection of slightly shallower depths with low food. Under high perceived risk of fish predation, a decrease in food availability

resulted in *Daphnia* selecting shallower depths. However, this change in
elevation was small and occurred only in the daytime.

   **Pattern 3. Response of reproductive allocation to predation
risk.** Under low or absent perceived fish predation risk, the *Daphnia* in
Fiksen's model initially allocated almost all their assimilated mass to
growth instead of reproduction. This allocation let them reach maximum
size quickly and switch to high reproductive output. At higher risk, the
Daphnia allocated some mass to reproduction from the start, producing
some offspring before reaching maximum size. However, at very high
risk the *Daphnia* fed very little, grew slowly, and allocated only a moderate fraction of mass to reproduction. We expect the optimization result
under very high risk to be somewhat arbitrary because the probability of
surviving until the time horizon is extremely low under any behavior; the
optimal behavior yielded a negative intrinsic rate of increase.

2. **Entities, state variables, and scales**. The model, like Fiksen's, represents
   a 1m water column occupied by *Daphnia*. The column is discretized into
   10 vertical layers of equal height (figure 5.2). We refer to these layers as
   patches, as they are analogous to the patches in the forager model of chapter
   4. Patch state variables include depth D (m, distance from the water surface
   to the bottom of the patch), temperature T (°C), and irradiance I (μmol/m²s),
   which is a measure of light intensity. T is static over a simulation, while I
   switches between day and night values and light decreases with depth. Fish
   concentration P (fish/L) is constant over both space and time.

   Daphnia are represented as individuals that can move among patches,
   with their location defined by which patch they are in. *Daphnia* have
   dynamic state variables for somatic body mass (X, μg dry mass), mass
   allocated to reproduction (R, μg dry mass), and number of offspring

FIGURE 5.2. Patches in the vertical migration model, labeled
by depth D (m) and shaded by irradiance I during daytime.

produced (N, an integer). A fourth state variable α is the fraction of growth allocated to reproduction; α varies over time and can have one of six discrete values, from 0 to 1 by 0.2.

The model uses a 1-hour time step, and simulations run for 1000 hours (41⅔ days).

3. **Process overview and scheduling**. The following actions are executed each time step. The actions use submodels that are fully described in element 7 below.

   a. *Environment updates* advance the time by 1 hour and update the patch irradiance I on time steps when conditions switch between day and night (via the light submodel, below). Daytime light conditions start at hour 4 and night conditions start at hour 20.

   b. *Habitat selection and reproductive allocation* combines the adaptive decisions by *Daphnia* of whether to move up or down, and how far, and how much energy to allocate to reproduction and growth. All *Daphnia* execute this submodel, in randomized order.

   c. *Growth* is executed by all *Daphnia*, in arbitrary order. The growth submodel updates the individual's size variables X and R.

   d. *Reproduction* is also executed by all *Daphnia*, in arbitrary order. The reproduction submodel determines how many offspring each *Daphnia* produces and updates N and R accordingly.

   e. *Predation survival* determines whether each individual *Daphnia* survives two sources of predation risk: invertebrate predators and fish. This action is executed by all *Daphnia*, in arbitrary order. Survival is a stochastic event: the *Daphnia* survives with a probability that combines the two kinds of predation risk. If the *Daphnia* does not survive, it is immediately removed from the model. (In many of the analyses below, the survival action is deactivated so the individual being simulated never dies. Instead, the action is modified to update the cumulative probability of survival from the start of the simulation.)

4. **Design concepts.**

   a. *Basic principles*: This model addresses the basic concept that diurnal VM results from an adaptive trade-off between food intake and predation risk driven by light level. This trade-off has been modeled for many species and in different ways; here we adopt the concept of Fiksen (1997) that the trade-off is between the fitness imperatives of avoiding fish predation and producing offspring.

   b. *Emergence*: The model predicts the vertical location of *Daphnia* over time, their growth and reproductive rates, and the abundance and biomass of a *Daphnia* population under laboratory conditions. These outcomes emerge from the vertical pattern of temperature and predation risk, light availability and how it differs between day and night, and the adaptive decisions of individuals.

c. *Adaptation*: Model *Daphnia* make a pair of contingent adaptive decisions each simulated hour: whether to move up or down, and how to allocate the mass they accumulate to either somatic growth or reproduction. Individuals make these decisions by evaluating a fitness measure for each combination of three to five possible patches and six reproductive allocation values and selecting the combination that provides the highest value of the fitness measure.

d. *Objectives*: We evaluate three alternative objective functions—fitness measures—used in the adaptive decisions. Sections 5.3–5.5 describe these alternatives.

e. *Learning*: No learning is assumed.

f. *Prediction*: The model *Daphnia* evaluate their fitness measures using explicit predictions of future predation risk, growth, and (in two alternatives) size and reproductive output. Sections 5.3–5.5 describe the assumptions in the predictions.

g. *Sensing*: The individuals can sense environmental conditions up to two patches away, without error.

h. *Interaction*: Model *Daphnia* do not interact.

i. *Stochasticity*: The only stochastic component of the model is in the predation survival submodel: predation survival probability is a deterministic function of *Daphnia* and habitat variables, but whether each individual survives each time step is a random event.

j. *Collectives*: There are no collectives.

k. *Observation*: While the model produces output for each 1-hour time step, we summarize results to make them comparable to the three observed patterns identified above and the results of Fiksen (1997) illustrated in figure 5.1. We graph VM results at 3-hour intervals. Reproductive allocation results are presented as daily mean α values to illustrate long-term trends instead of within-day variation. To test the model's ability to reproduce the second pattern, we simulated Fiksen's high and low food scenarios by changing light and growth parameters as described in element 7 below.

5. **Initialization**. The model is initialized at midday (12:00). The light availability in each patch is set using the light submodel. Patch temperatures T are set to values determined from figure 1 of Fiksen (1997): for depths 0.1 to 1.0, patch temperatures are 20.2, 19.6, 18.8, 16.2, 13.0, 12.0, 11.3, 10.9, 10.2, and 10.0°C. These conditions produce the profiles of temperature and growth illustrated in figure 5.3.

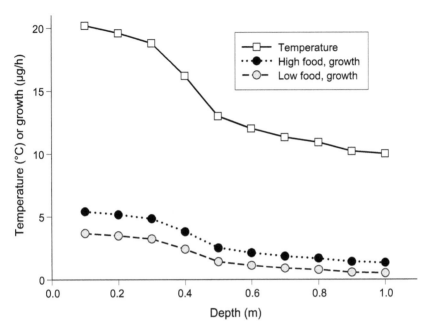

FIGURE 5.3. Profiles of temperature and growth with depth. Growth is the increase in mass (μg) per hour for a 200 μg *Daphnia* at the two food scenarios.

   The number of *Daphnia* created at the start of a simulation varies among the simulation experiments. *Daphnia* initially have a mass X of 21.5 μg, reproductive mass R of zero, and no offspring N. The *Daphnia* are placed in the highest patch.

6. **Input data**. The model uses no time-series input.
7. **Submodels**. The model actions (element 3) include five submodels described in detail here.

   The *light submodel* calculates irradiance I in each patch, as a function of surface irradiance and depth. It uses Beer's law, the standard equation for light attenuation in a turbid medium: $I_z = I_0 e^{-zK}$, where I_z is the irradiance at depth z (m), I_0 is the irradiance just below the surface, and K is the light attenuation coefficient, estimated by Fiksen (1997) as 2.6 m^{-1} for the high food scenario and 1.28 for the low food scenario (the phytoplankton that *Daphnia* eat are major contributors to light attenuation). The values of I_0 representing day and night are 17.9 and 0.1 μmol/m^{2}s, respectively. The value of z for each patch is the depth of its midpoint (0.05 m for patch 1, 0.15 m for patch 2, etc.).

   The *habitat selection and reproductive allocation submodel* represents the adaptive behaviors that determine which patch *Daphnia* move to and what value of α they select for the current time step. The *Daphnia* make these decisions by calculating the value of a fitness measure for all combinations of (a) up to five patches, those within 20 cm above and below their current depth, and (b) all six potential values of α. In sections 5.3 through 5.5 we define and evaluate alternative versions of the fitness measure and predictions used to rank these combinations of patch and α.

The *growth submodel* represents how *Daphnia* add mass and allocate it between size and reproduction. Growth (G, µg dry mass) during a 1-hour time step is modeled using an empirical equation that assumes growth is controlled only by *Daphnia* size and temperature: $G = X[\exp(\frac{\rho T - 0.31}{24}) - 1]$, where ρ is 0.047 for high food availability simulations and 0.037 for low food availability simulations. The *Daphnia* variable α determines how much of G is allocated to increase in size vs. reproduction. The *Daphnia*'s reproductive mass R is increased by αG and its size X is increased by $(1 - \alpha)G$. However, this submodel enforces a maximum size: if the new value of X exceeds 400 µg, it is reduced to 400.

The *reproduction submodel* calculates the number of offspring produced. This submodel is simply a way to track an individual's fitness; new offspring are not actually created and added to the model. The submodel simply assumes that producing one offspring requires 6 µg of mass. Therefore, the values of reproductive mass R and number of offspring N are updated by (a) increasing N by the value of $R/6$ truncated to an integer (which is often zero), and (b) resetting R to the remainder of $R/6$.

The *predation survival submodel* determines the probability of surviving two kinds of predation and how survival varies with *Daphnia* and habitat characteristics. Risk from invertebrate predators is assumed to be independent of light because these predators do not rely primarily on sight to find prey; and larger *Daphnia* are assumed to be less vulnerable to invertebrate predators. Parts of Fiksen's predation submodel were not completely described; we approximated it for survival of invertebrates (S_I, probability of surviving for one day) as $S_I = 1 - 0.613X^{-0.24}$.

The probability of surviving predation by fish is assumed to decrease with both *Daphnia* size and irradiance because both make *Daphnia* easier for fish to see. Fiksen used the cross-sectional area of a *Daphnia* as the size measure determining the risk of fish predation; using Fiksen's assumptions, this area A (cm^2) = $10^{-4}X^{2/3}$. We approximated Fiksen's fish predation function as $S_F = \exp(-2600I_zPA)$, where S_F is the probability of surviving fish predation for one day and I_z and P are the irradiance and fish concentration as defined above.

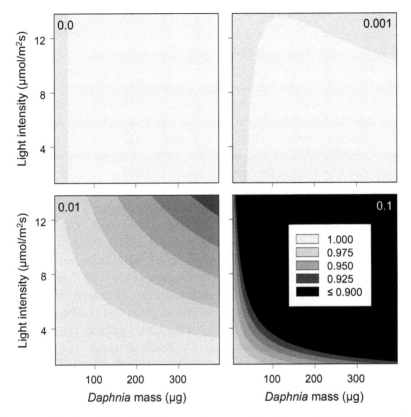

FIGURE 5.4. Hourly survival probabilities over the ranges of irradiance and *Daphnia* mass occurring in the model. Probabilities are shown for four fish concentrations: 0.0, 0.001, 0.01, and 0.1 fish/L.

The probability S of surviving both kinds of predation for one hourly time step is calculated as $S = (S_I S_F)^{1/24}$. The two predation functions produce rather complex and nonlinear relations among *Daphnia* size, irradiance, fish concentration, and survival probability (figure 5.4).

5.3 SPT VERSION 1: EXPECTED FUTURE REPRODUCTION WITH CURRENT GROWTH AND SURVIVAL

Now let us look at how SPT can be applied to the contingent adaptive decisions of our *Daphnia*: whether to move up or down, and how much energy to allocate to growth vs. reproduction. The *Daphnia* do this by evaluating the available combinations of patch and α and selecting the one that provides the highest value of a fitness measure. Our job now is to define a fitness measure and the predictions used by the model *Daphnia* to evaluate it.

The first version of SPT we develop is the simplest. Our fitness measure represents expected reproductive potential at the time horizon, approximated as the accumulated reproductive mass R at the time horizon times the probability of surviving until the time horizon. The simulated *Daphnia* calculate expected fitness not by predicting how many offspring they would produce but by simply predicting the total mass allocated to reproduction. Representing reproductive output as accumulated R ignores all the details of when offspring are reproduced and how likely they are to survive and reproduce. (Fiksen [1997] discusses and contrasts alternative ways to represent reproduction in a fitness measure.)

This version also uses a simple prediction: we assume the *Daphnia* predict that the growth rate (μg/h) and survival probability they would experience on the current time step, for any combination of patch and α value, will not change until the time horizon. This prediction is not accurate even in the *Daphnia* model, where most environmental variables remain constant over time. Both growth and survival depend on *Daphnia* mass and therefore change as *Daphnia* grow even if nothing else changes. But in this version the fitness measure ignores changes in growth and survival due to both growth and diurnal changes in light. Assuming that *Daphnia* do not predict the daily light cycle in their behavior may seem extremely unrealistic, but we nevertheless want to test how well we can model their behavior under that assumption.

Implementing these assumptions, the fitness measure $F = (G\alpha h)(S^h)$, where G and S are the growth and survival that would be obtained in the patch on the current hour and h is the number of hours until the time horizon. The first term extrapolates the current accumulation of reproductive mass to the time horizon, and the second term is the probability of surviving to the horizon.

Now we must define the time horizon. Fiksen's original model used a time horizon of the 1000th hourly time step, because the dynamic programming optimization used by DSVM requires a fixed time horizon. One thousand hours is probably longer than necessary for modeling an individual's behavior because the probability that a *Daphnia* could survive this long, even under the safest conditions available (mass of 400 μg and zero fish or zero light) is about one in one thousand. On the other hand, a very short time horizon also produces unrealistic behavior because it undervalues the importance of survival. If we run the *Daphnia* model assuming a very short time horizon, the *Daphnia* never exhibit VM: survival over a few hours is too high to cause them to leave the surface level where growth is highest.

Fiksen (1997) noted that a fixed time horizon causes "terminal effects": behavior changes near the end of a simulation because survival becomes less important as the time left to survive decreases. Such effects may be realistic when the time horizon represents a major life history event, such as transitioning to a completely different life stage or semelparous reproduction, but is not realistic for the *Daphnia* model with its single life stage and continuous reproduction.

Considering these factors, we chose a "sliding" time horizon of 10 days: on each hour, the *Daphnia* evaluate their fitness measure over 240 future hours; h in the fitness measure equation is always 240. The sliding time horizon avoids terminal effects, and at 240 h it gives full weight to survival: even under the safest conditions possible, a *Daphnia* has only a 20% probability of surviving to the horizon.

Now, how well does this version of SPT cause the *Daphnia* model to reproduce the patterns identified above (section 5.2)? We reproduced Fiksen's "dilution experiment," simulating eight fish densities ranging from 0 to 0.2 fish per liter under the high food scenario. For comparison to laboratory data and Fiksen's results in figure 5.1, we plot *Daphnia* location every 3 hours during the first 96 simulated hours. Because we are looking at behavior of an individual *Daphnia*, with no interaction among individuals and no stochasticity in the model, we need only simulate one individual and turn off the predation action so it does not die.

Even this very simple implementation of SPT produces VM surprisingly similar to that observed in the laboratory and produced by Fiksen's DSVM model (figure 5.5). It produces the basic pattern of no migration at low fish concentrations, strong diurnal migration at intermediate concentrations, and (better than the DSVM model) complete avoidance of shallow depths at the highest fish concentrations. It also reproduces the pattern of slightly more use of shallow depths with lower food availability (figure 5.6). The one VM characteristic not reproduced is *Daphnia* starting migration only after a period of growth at the lowest fish concentration that produces migration.

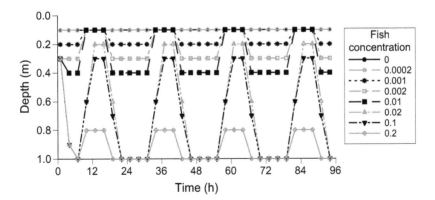

FIGURE 5.5. Vertical migration with the first version of SPT, for comparison to figure 5.1.

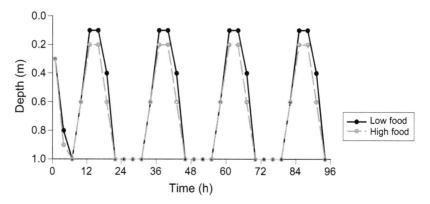

FIGURE 5.6. Vertical migration in version 1, at two food availability levels and fish concentration of 0.02. The low and high food scenarios are defined in the light and growth submodels.

However, this version did not reproduce the patterns in allocation of mass to reproduction and growth produced by the DSVM model. Instead, it predicted that *Daphnia* always select $\alpha = 1.0$, so they contribute all their food intake to reproduction and never grow. The failure to reproduce *Daphnia* growth indicates that this simplest version of SPT is missing important mechanisms: the fitness benefits of current growth. Predicting that current survival probability and rate of reproductive mass accumulation persist over the time horizon provides no consideration of how growth increases survival (when in dark habitat), future size, and the future rate of offspring production. So let us see next whether we can represent such mechanisms via SPT, and if so, how much more complex the fitness measure needs to become.

5.4 SPT VERSION 2: PREDICTED OFFSPRING

Our first version of SPT could produce VM but failed to produce reasonable reproductive allocation because it did not consider the future fitness benefits of current growth. Now we try to remedy this problem while still keeping the behavior submodel as simple as we can. We therefore still assume that the *Daphnia* ignore future diurnal light cycles and still predict constant habitat conditions. But now our virtual *Daphnia* no longer assume that their growth and reproductive mass accumulation rates and survival probabilities remain unchanged until the time horizon.

How could we model an individual *Daphnia*'s expected fitness at the time horizon in a way that captures the effect of current growth on future reproductive rate? The assumption we adopt from Fiksen (1997) that size is limited to 400 μg further complicates projection of future size, growth, and reproduction.

We try an approach that still assumes *Daphnia* predict their future fitness for a combination of patch and α by assuming they will use the same patch and α value until the time horizon (with one exception). We define fitness at the time horizon as the expected number of offspring; specifically, the predicted number of offspring produced before the time horizon times the predicted probability of surviving until the time horizon. We will try to predict future offspring production and survival probability fairly accurately even as they vary nonlinearly as *Daphnia* grow.

With *Daphnia* predicting that their future environmental conditions will be the same as at their current patch, we could predict growth trajectories and reproduction quite closely. Because growth is a linear function of X, we could use the exponential growth equation to approximate the time at which a *Daphnia* reaches the maximum mass of 400 μg. (The only reason the exponential growth equation is an approximation is because it represents time as continuous, while our model—like most IBMs—uses discrete time steps.) With a growth trajectory calculated this way and a constant value of α, we could easily calculate energy allocated to mass increase and to reproduction, and therefore approximate the number of offspring produced. Unfortunately, predicting future *survival* mathematically is not as easy because survival varies nonlinearly with size.

Instead of using mathematics to predict future reproduction and survival, we can use an algorithmic approach, iterating forward through future hours until the time horizon and updating expected size, offspring, and survival for each predicted time step. Our version 2 *Daphnia* use these steps to evaluate combinations of patch and α value:

1. Set predicted survival until the time horizon S_p to 1.0, predicted mass X_p to the current mass X, predicted reproductive mass R_p to the current value of R, and predicted number of offspring N_p to N.
2. Repeat the following steps once for each of the 240 hours until the time horizon:
   a. Calculate the hourly growth G_h that would be obtained in the patch with mass X_p.
   b. Increment R_p by $αG_h$ and increment X_p by $(1 - α)G_h$. However, if X_p reaches the maximum mass of 400 μg, then X_p remains at 400 and the value of α is changed to 1.0 for this and all remaining predicted hours. The value of α is switched to 1.0 because once maximum mass is reached the optimal strategy must be to allocate all intake to reproduction.
   c. Simulate reproduction, updating N_p and R_p using the growth submodel.
   d. Calculate the survival S_h that would be obtained in the patch with mass X_p. Update S_p by multiplying it by S_h.
3. Calculate expected fitness as $N_p S_p$.

In step 2.b of this algorithm, we introduced an assumption that the value of α switches to 1.0 when a *Daphnia* reaches maximum mass. This assumption, which turns out to be critical for producing the life history patterns discussed below, is also implemented in the growth action that determines the actual (not predicted) value of *R* each time step.

How does the SPT model perform now that we represent how growth affects future survival and reproduction? Version 2 clearly reproduces the basic VM patterns (figure 5.7). The difference between low and high food scenarios is very similar to that in version 1. However, unlike version 1, the model now reproduces the pattern of *Daphnia* making less use of shallow depths as they grow.

Also unlike our first attempt, version 2 produces distinct patterns in the life history variable α (figure 5.8). As in Fiksen's DSVM model, low fish predation

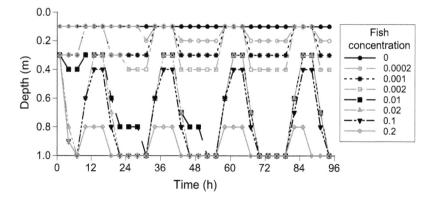

FIGURE 5.7. Vertical migration in version 2 of the *Daphnia* model, high food scenario.

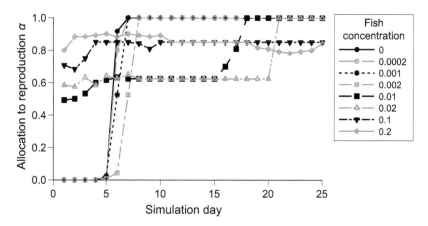

FIGURE 5.8. Daily mean α values selected by *Daphnia* in version 2, high food scenario.

risk causes *Daphnia* to initially allocate mass to growth, allowing them to reach maximum size and reproductive output rapidly. High risk causes more allocation to reproduction from the start of the simulation.

5.5 SPT VERSION 3: DIURNAL PREDICTION

Even though our first two SPT models reproduce most of the patterns that characterize the *Daphnia* VM system, we wonder whether their performance could be improved by eliminating the clearly wrong prediction that future habitat conditions remain the same as those in the current time step. Specifically, it seems very likely that real *Daphnia* are well adapted to diurnal changes in light levels, as those variations both strongly affect predation risk and are very predictable. It seems likely, therefore, that our model *Daphnia* will have higher fitness if we let them consider the diurnal cycle in their adaptive behavior.

In this version we modify the algorithm for predicting future reproduction and survival to include differences between day and night. We do this simply by keeping track of whether each of the future hours evaluated (step 2 of the algorithm described in section 5.4) is in day or night and using memory of habitat conditions during the previous phase of the day-night cycle to predict future conditions. A *Daphnia* deciding what to do during a daytime hour uses the prediction that future daytime hours until the time horizon will all have the same growth and survival conditions as the patch it is evaluating, and that future nighttime hours will have conditions it "remembers" from the patch and α value the *Daphnia* actually used during the most recent night hour. Now the *Daphnia* evaluates a particular combination of patch and α value by assuming it would use one combination during the day and another combination at night.

Giving *Daphnia* the ability to consider diurnal variation in risk produced only small changes in their behavior (figure 5.9, compared to figure 5.7). At intermediate levels of fish predation risk, *Daphnia* in version 3 occupied slightly deeper habitat at some times; "knowing" that they could feed safely at night let them reduce their daytime risk.

The differences in reproductive allocation were less subtle (figure 5.10, compared to figure 5.8): with consideration of diurnal change in risk, *Daphnia* at intermediate predation levels allocated more mass to growth early in the simulation and attained maximum size sooner, while *Daphnia* exposed to the two highest predation risk levels allocated all their mass to reproduction and never grew in size. This version again captures the major trends in Fiksen's (1997) optimized values of α, but Fiksen did not provide sufficient detail to allow us to determine which of our SPT versions is closest to his DSVM results.

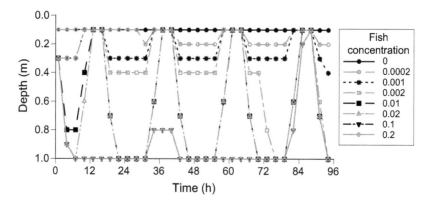

FIGURE 5.9. Vertical migration in version 3, high food scenario.

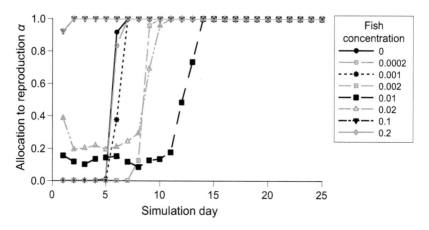

FIGURE 5.10. Daily mean α values in version 3.

5.6 PREDICTION COMPLEXITY AND FITNESS:
POPULATION SIMULATIONS

So far in this chapter we have shown that three versions of SPT with increasing complexity can all reproduce the basic patterns of VM observed in the laboratory and produced by Fiksen's (1997) DSVM model. But we also see that the different versions produce minor differences in VM and sometimes strong differences in reproductive allocation. We have not yet, though, answered the questions of which version of SPT is "best" and how *Daphnia* fitness changes with the complexity of the behavior submodel.

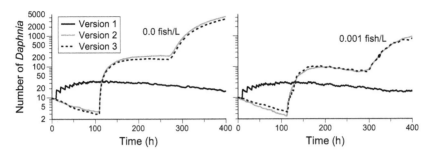

FIGURE 5.11. Population simulation results: mean abundance of *Daphnia* over 10 replicate simulations, with fish concentrations of 0.0 and 0.001 fish/L.

Because we are using SPT, we can easily turn our model of an individual *Daphnia* into a simple population model that includes reproduction and mortality over generations, though we choose not to introduce interactions among individuals. Simulations were initialized with 10 *Daphnia*. We modified the reproduction submodel so it actually creates new *Daphnia* whenever an existing *Daphnia* reproduces. New *Daphnia* are created with $X = 6$ μg and $R = 0$, and placed in the patch where their parent reproduced. We included mortality by activating the predation submodel. We conducted simulations using all three versions of SPT and executed 10 replicates of each fish predation level under the high food scenario.

This population model contains no internal feedbacks to regulate it, so it unsurprisingly produced either rapid population growth or rapid decline depending on the risk of fish predation. At zero and low fish predation, population growth was so rapid that we stopped the simulations at 400 h. The differences between the SPT versions with and without prediction of the future benefits of growth are clear (figure 5.11). Version 1, which does not predict future benefits to reproduction of current growth, produces near-steady abundance as *Daphnia* contribute all their growth to immediate reproduction. In contrast, the other two versions delay reproduction until *Daphnia* have achieved maximum size, producing two distinct generations of offspring over the 400 hours. Versions 2 and 3 produce similar results, indicating that the additional complexity of representing day and night in predicting future fitness has little or no payoff at the population level.

High levels of fish predation caused our simulated *Daphnia* population to decline rapidly, so that we had to increase the initial number of *Daphnia* to 100 to keep the population from being extirpated almost immediately. At fish concentrations of 0.01, the population persisted for 1000 h, but at 0.1 fish/L it was extirpated within 400 h (figure 5.12). In these simulations, the first SPT version again performed better early in the simulations. Because high predation risk causes

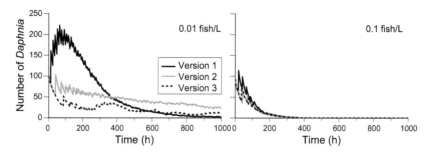

FIGURE 5.12. Population simulation results with fish concentrations of 0.01 and 0.1 fish/L, with 100 initial *Daphnia*.

Daphnia to reduce feeding by staying in deep habitat, at 0.01 fish/L it took almost 400 h for the SPT version 2 population to exceed that of version 1, and version 3 performed distinctly worse than version 2.

5.7 CONCLUSIONS

Our main objective in this chapter was to illustrate the application of SPT to a real biological system, but a system that has been simplified enough to also be addressed by DSVM. The *Daphnia* VM system let us explore ways to use SPT for an adaptive trade-off behavior that clearly is important to population dynamics, and to compare the behavior it produces to behavior predicted by a DSVM model and observed in real organisms. We succeeded in reproducing the three patterns of observed behavior, in almost all their details, while converting Fiksen's (1997) DSVM model into a population model that includes mortality and reproduction and could easily include processes that make DSVM intractable—like interaction and variation among individuals.

One lesson from this chapter is that SPT can readily represent contingent decisions, such as *Daphnia* choice of both depth and reproductive allocation. This is not surprising because such decisions have also been modeled with DSVM. The approach we adopted from Fiksen (1997)—and used in the trout model of chapter 2—is simply to evaluate the fitness measure for combinations of the two decision variables. The problem is that examining combinations of alternatives dramatically increases computational effort. By simplifying the fitness measure and eliminating the need for dynamic programming, SPT can make contingent decisions more tractable. In chapter 7 we illustrate a much more complex contingent decision: modifying the trout model's habitat selection decision to include an irreversible life history decision.

The *Daphnia* model has one interesting difference from the trout model and the patch selection model of chapter 4. The primary decision in all these models is foraging habitat selection; but the *Daphnia* model treats this decision as a trade-off between predation and reproduction, while the other models treat it as a trade-off between predation and starvation. Consequently, the SPT design was focused on predicting future reproductive potential instead of future starvation. How to represent the fitness benefits of future reproduction, e.g., whether to simply predict the number of direct offspring, as we did, or to factor in the reproduction and survival of those offspring, as Fiksen (1997) did, is an issue we return to in chapter 8.

One reason that we chose not to further explore "details" such as how to represent fitness effects of future reproduction is that the VM behavior seems quite easy to reproduce and robust to model details. This conclusion is not surprising because VM is driven, at least in this model, by very strong gradients in growth and predation risk. Even our simplest, most approximate version of SPT produced most characteristics of the observed VM.

The patterns in reproductive allocation produced by our model varied more among the versions of SPT and did not always closely match those produced by Fiksen's DSVM analysis, especially at very high predation levels. These discrepancies could be caused by differences in how future reproduction was represented in the two approaches. Under the extreme predation levels that produced the biggest discrepancies, all the behavior models are probably especially sensitive to method details. However, at these predation levels the *Daphnia* were doomed anyway; exactly how to model individuals with ~50% probability of surviving a single day is unlikely to be important for population ecology.

We think several methodological lessons from this chapter have general value. First, in calculating future fitness consequences of the reproductive allocation decision, we found it simpler to use an algorithmic approach instead of equations. We could predict future size, offspring production, and survival probability by stepping forward through time. Calculating survival probability over time is often especially difficult because per-time-step probabilities tend to vary nonlinearly with characteristics of individuals and their habitat and must be multiplied, not added, together. We can rely on algorithmic simulation as well as approximation when it is not practical to use math alone to represent expected future fitness. In this case, using the forward time step algorithm to calculate expected future fitness slowed the model's execution but did not make it cumbersome until the number of individuals reached the thousands.

We also saw that a sliding time horizon, in which individuals always calculate expected future fitness over the same number of future time steps, has advantages over the fixed time horizon that DSVM typically uses. In some ecological situations, a population cohort may actually have something like a fixed time horizon, such as a date or size at which a life history transition must be made. But very often there is no such event and using a fixed time horizon can introduce the kinds of "terminal effects" that Fiksen noted in his DSVM model: behavior changes only because the time horizon is being approached and survival until the horizon is less important. Another benefit of a sliding time horizon is that it lets us introduce individuals at different times—here, via *Daphnia* reproduction—without their behavior being affected by when they were created. In our next chapter we try an alternative to both fixed and sliding time horizons. Then in chapter 7 we model a life history decision with a sliding time horizon that makes the date of transition between life stages not a fixed time horizon that must be defined a priori but an outcome of each individual's experience.

Finally, we observed that making the fitness measure and prediction assumptions more complex did not always make the simulated adaptive behavior more successful. When we modified our behavior model so that the *Daphnia* consider the diurnal differences in growth and survival, the resulting version 3 produced population growth that was no better, and sometimes worse, than the simpler version 2. This example again illustrates the challenge of identifying reasonable levels and kinds of complexity in the behavior component of population models; we address this challenge in chapter 10.

CHAPTER 6

Example Three: Temporal Patterns in Limpet Foraging

6.1 BACKGROUND AND OBJECTIVES

In this chapter we present a third example application of SPT, again involving a behavior originally modeled via DSVM. This example also addresses animal foraging, this time a choice of foraging activity. In this case, physiology has more important and interesting effects on behavior: the model animal, like many herbivores, can consume food relatively rapidly but does not assimilate its energy until the food's rather slow passage through a long gut. This leads to uncoupling of foraging behavior and energy assimilation on short time scales.

The example is based directly on the DSVM analysis by Santini et al. (2014) of foraging behavior in the limpet *Cella grata*, which feeds by scraping rocks in the intertidal zone (Santini et al. 2011). Staying in the wave splash during favorable tides protects limpets from both predation and desiccation: the splash conceals the limpets and discourages predators from approaching while providing the moisture limpets need to move and graze. During unfavorable wave conditions, usually at low and high tide, the limpets move up the shore and rest until favorable wave conditions return (e.g., when rising tide brings the waves to their resting site). At some times during favorable feeding conditions, the limpets cease feeding activity but remain in the wave zone instead of expending the energy to move to the resting zone (Santini et al. 2011). The limpets' foraging problem therefore appears to be to feed long enough to fill their gut often enough, and at the right times, to maintain energy reserves and meet the cost of moving daily to a resting site.

In this chapter we briefly present the DSVM model of Santini et al. (2014), then fully describe the model we built for the same behavior. We use SPT to model the adaptive behavior of an individual, not a population of interacting individuals, but in a way that could be used in a population-level IBM. We present and explore several SPT formulations, comparing their results to those of the DSVM and to field observations of *Cella grata* by Santini et al. (2011). However, the original model of Santini et al. is highly simplified, e.g., by representing only one tide cycle per day and not representing space explicitly, so the models cannot be contrasted by how closely they reproduce quantitative field observations.

Online models and NetLogo computer code for all limpet models presented here are available on this book's web site.

6.2 THE DSVM MODEL OF LIMPET FORAGING

The model of Santini et al. (2014) represents foraging as the choice over time of three activities: resting in safe habitat with no mortality risk and no food intake, feeding in foraging habitat, and *stasis*: resting in the foraging habitat instead of making the costly return to resting habitat. Daily tidal cycles are represented very simply by dividing each day into two periods. The "favorable" period represents intermediate tides when wave splash protects feeding limpets, and the "unfavorable" period represents tides when limpets would be exposed to predation and desiccation if they fed. The transition from feeding or stasis to resting incurs an additional energy cost of movement. The DSVM represents mortality from predation and desiccation in a binary way: survival is zero if a limpet feeds or uses stasis during the unfavorable period but one otherwise.

This model is interesting because energy assimilation during a (½-hour) time step results from digestion of food consumed during earlier time steps. Model limpets have separate state variables for energy reserves and the volume of unprocessed food in their guts. Energy assimilation is limited not by feeding rate but by the processing rate of food in the gut, and the gut can store enough food to be processed over several time steps. Hence a limpet's energetic state depends more on its foraging behavior in previous time steps than in the current one.

Santini et al. (2014) formulated their model as a DSVM optimization with energy reserves at the end of an 8-day period as the fitness measure: the optimization selects the activity of a limpet on each half-hour time step to maximize its energy reserves at the end of 384 steps. For several reasons, many combinations of activity over time can produce exactly the same terminal energy reserves without incurring mortality, so behavior through much of the 8-day period is partially arbitrary. First, the model's simple representation of digestive physiology and energetics uses parameters that set maximum gut content and processing rates and a maximum energy reserve; consequently, many combinations of activity over time can bring energy reserves up to their maximum. Second, because a limpet can completely replenish its gut contents and reach the maximum energy reserves in a few hours, only behavior on the final day affects energy reserves at the 8-day time horizon, and feeding decisions earlier in the simulation have no effect on the fitness measure as long as mortality is avoided. Finally, the optimization results are partially arbitrary because the model lacks complexities such as variation over space or time in feeding rate and mortality risks other than zero or one. Because of this arbitrariness, we give less emphasis to how our SPT results compare to the DSVM optimization and more to how successfully the alternative SPT models reproduce behaviors of real limpets that Santini et al. (2014) used to evaluate their model.

6.3 THE MODEL

Our model closely follows the DSVM model of Santini et al. (2014); the only intentional differences are in how we modeled behavior. Here we describe the model in the ODD format, except that we describe and analyze two alternatives for modeling the foraging decision in separate subsections. We describe the model as an IBM even though we use only one limpet in our analysis and the model does not include differences or interactions among individuals.

1. **Purpose and patterns**. Santini et al. (2014) designed their model to over-come the limitations of static energy maximization theory for explaining foraging behavior, especially the inability of that theory to represent temporal effects of physiological and habitat processes. Their model's main purpose seems to be explaining and reproducing a set of patterns observed in field experiments of Santini et al. (2011). These patterns focus on three different parts of the relation between the mean number of hours per day that limpets spent in the wave zone (H_A) versus the number of hours per day of tidal conditions favorable for feeding (H_F): (1) when H_F was below about 5 hours per day, H_A equaled or even exceeded H_F; (2) over the central range of H_F (about 5 to 15 hours per day), limpets were active almost throughout the favorable period; and (3) when H_F exceeded about 15 hours, H_A no longer increased with H_F—additional hours of favorable conditions did not produce more activity.

2. **Entities, state variables, and scales**. The only entity explicitly represented is one individual limpet. Its environment is represented only via a true-false variable *favorable?* that is true when tidal conditions favor feeding. An additional variable represents the time (0 to 23.5 hours) at the beginning of the current time step.

   The limpet has dynamic state variables w for its energy reserves (J); g for gut contents, the volume (mm^3) of food ingested but not yet processed into energy; and *activity*, which uses three possible values to describe current behavior. The value of *activity* is "feeding" when a limpet has chosen to feed, "stasis" when not feeding but remaining in the wave zone, and "resting" when the limpet has stopped feeding and moved out of the wave zone. (Other limpet characteristics such as size are incorporated in model parameters because they do not change.)

Space is not represented explicitly. Time is represented via ½-hour time steps, and simulations cover 8 days.

3. **Process overview and scheduling**. The model executes the following actions in each time step. Two actions use submodels that are fully described in element 7 below.

  a. *Environment updates* revises the model time and the value of *favorable?*. One half hour is added to the model time; a value of 24 resets to 0. The model uses a highly simplified representation of the complex tidal cycles that actually drive limpet foraging. Tidal conditions are represented only via the parameter T_{PAP}, the number of hours per day of favorable feeding conditions (the "potential activity period"). The favorable conditions are assumed to start at hour 0 of each day and continue until the model time equals T_{PAP}, at which time *favorable?* is set to false.

  b. *Activity selection* determines which of the three activities the limpet selects, using one of the alternative submodels described in sections 6.4 and 6.5.

  c. *Energetics* updates the limpet's energy reserves w and gut contents g, by executing the energetics submodel described below.

  d. *Survival* determines whether the limpet survives mortality that represents (a) starvation and (b) risks such as predation and desiccation. The SPT models use the simple, binary representation of survival used in the original DSVM. A limpet always survives starvation if its value of w is positive and dies if not. It always survives predation and desiccation if its activity is "resting" or if *favorable?* is true, and always dies if its activity is "feeding" or "stasis" when *favorable?* is false.

4. **Design concepts.**

  a. *Basic principles*: The original DSVM model's basic principle was the use of dynamic simulation with trade-off decisions to explain real limpet behaviors that could not be explained by the static energy maximization theory used in previous models.

  b. *Emergence*: The primary model result of interest is the daily pattern of foraging activity selected by the model limpet, specifically the number of hours of activity per day. This pattern emerges from the limpet activity selection behavior and the number of hours per day of favorable conditions (H_F).

  c. *Adaptation*: Limpets make one adaptive decision—selecting among the three alternative foraging activities. This decision is made by selecting the alternative providing the highest value of a fitness measure, although the decision also depends strongly on rules for which alternative to choose in the common situation that multiple activities offer the same fitness value. Sections 6.4 and 6.5 specify these rules.

d. *Objectives*: The decision objective, or fitness measure, represents expected energy reserves w at a future time horizon, the product of predicted energy reserves and probability of survival to the time horizon. Survival depends on both predation and starvation, so this objective represents a trade-off. Sections 6.4 and 6.5 describe and explore alternative fitness measures and time horizons.

e. *Learning*: No learning is assumed.

f. *Prediction*: The fitness measures assume that limpets predict gut contents and energy reserves, and survival probability, between the current time and the time horizon. Sections 6.4 and 6.5 describe prediction methods.

g. *Sensing*: The limpets are assumed to know, for activity selection, their current gut contents g and energy reserves w, and their probability of surviving each activity during favorable and unfavorable periods; they are also assumed to know the value of H_F and the current time. (Santini et al. [2011] cite Gray and Williams [2010] as evidence that *C. grata* has an internal clock that allows them to estimate tidal cycles, supporting the assumption that the model limpets know the favorable period length.)

h. *Interaction*: There is no interaction among limpets.

i. *Stochasticity*: This model is unusual in that it lacks any stochasticity.

j. *Collectives*: Collectives are not represented.

k. *Observation*: The model produces output reporting the activity selection of the simulated limpet. For comparison with results in Santini et al. (2014), the model output "active hours per day" is the number of hours of feeding or stasis per day. Because the model produces the same behavior pattern on each day after the first, we simply report active hours per day from the final day.

5. **Initialization**. Simulations start with the time at hour zero and the variable *favorable?* set to true. The limpet is initialized with: *activity* set to "resting," energy reserves w set to the maximum reserve w_{max}, and gut contents (g) set to zero.

6. **Input data**. The model uses no time-series input.

7. **Submodels**. Two alternative activity selection submodels are described in sections 6.4 and 6.5. We describe the energetics submodel here. This submodel follows the energetics formulation of Santini et al. (2014) as closely as possible but uses fewer parameters because we assume constant limpet

size and time-step length, which makes parameters for size and time-step dependencies unnecessary.

The energetics submodel first updates the limpet's energy reserves using an energy balance of intake from digesting gut contents and energy costs:

$$w_t = w_{t-1} - C + (\delta\sigma) \text{ while } w_t \leq w_{max}$$

where w_t is the energy reserve (in J) for the current time step, C is the energy cost (J) for the current time step, δ is the energy content of food (1.66 J/mm^3), and σ is the volume (mm^3) of food processed from the gut into energy. Energy reserves are limited to a w_{max} of 231 J. If w_t reaches w_{max}, processing of gut contents is assumed to continue but the energy is lost.

The energy cost C depends on the limpet's activity and changes in activity; the costs are for respiration and production of the mucus limpets expend during movement. The cost formulation of Santini et al. (2014) can be reduced to the following: if *activity* is "resting" and was "resting" the previous time step, then $C = 6.23$ J; if *activity* is "resting" and was "feeding" or "stasis" on the previous time step (so costs include movement to a safe resting place), then $C = 135.7$; if *activity* is "feeding," then $C = 49.4$; and if *activity* is "stasis," then $C = 6.23$ J.

Gut contents determine the value of σ. If the volume of food in the gut g_{t-1} is greater than the maximum that can be processed per time step, then σ is set to that maximum processed; otherwise, σ is set to g_{t-1}. The maximum that can be processed per time step σ_{max} is 23.6 mm^3. Therefore, the maximum energy gain per time step is 39.2 J.

This formulation means that a limpet with a full gut actually loses 10 J per time step if it continues to feed, but assimilates 27 J per time step if it switches to stasis and 33 J per time step while resting. But on the time step when a limpet switches from feeding or stasis to resting, it loses 97 J. Feeding therefore sacrifices current energy assimilation to fill the gut for future assimilation.

The second part of the energetics submodel is updating g, via a material balance that includes feeding as well as processing gut contents into energy:

$$g_t = g_{t-1} + \lambda - \sigma \text{ while } g_t \leq g_{max}$$

where λ is the food ingestion rate and g_{max} is the maximum gut content of 207 mm^3. The value of λ is zero unless *activity* = "feeding," when $\lambda = 70.7$ mm^3 per time step.

It is important to understand the scheduling of energy versus gut content updates: the energy reserves at the end of a time step depend on gut content at the start of the time step, not after the gut is (or is not) refilled via feeding on the current time step.

6.4 SPT VERSION 1: MAXIMIZING SHORT-TERM EXPECTED ENERGY RESERVES

We begin with a version of SPT for the limpet model that is very simple, yet informative about fitness measures and time horizons. The model does not include reproduction, so the adaptive foraging decision is a trade-off between (a) energy intake and reserves and (b) survival of predation and desiccation. Therefore, we follow Santini et al. (2014) and use as the fitness measure expected energy reserves $= S_T \times w_T$, where S_T is the probability of surviving to a future time horizon T and w_T is energy reserves at T. However, starvation is represented by setting the fitness measure's value to zero if w is predicted to fall to or below zero before or at the time horizon.

Although Santini et al. used a time horizon of 8 days after the start of the simulation, the benefit of such a long horizon is unclear. No future life history event either provides a natural time horizon or demands accumulation of a certain amount of energy. In fact, the long time horizon allows energy reserves to be low throughout most of the simulation because the limpet can replenish them on the last day; this artifact seems unrealistic and undesirable. Further, survival is either zero or one, so the time horizon has no effect on S_T. Therefore, for this first version of SPT we use the simplest possible approach: a sliding time horizon at the end of the current ½-hour time step.

Our SPT formulation is thus to select the activity that provides the highest value of $S_t w_t$, where t refers to the end of the current time step. However, the model's limits on gut contents, gut processing rate, and energy reserves (g_{max}, σ_{max}, and w_{max}, respectively) make it common that two or more activities provide exactly the same value of this fitness measure. To avoid ambiguities in the decision, we use the following rules: (1) a model limpet changes its activity only if the new activity offers higher, not equal, expected fitness; and (2) if feeding and stasis offer equal expected fitness, limpets choose stasis if feeding would exceed the maximum gut content. These rules undoubtedly have strong effects on simulated behavior.

This formulation produces simple, unrealistic, and—upon reflection—predictable results. The model limpet never feeds and dies as soon as its energy reserves fall below zero during an unfavorable period. Why does the limpet never change its activity from initial resting to feeding as its energy reserves fall to zero? Over just one time step, switching to feeding increases metabolic costs and produces no energy assimilation: remember that the limpet assimilates energy not when it feeds but when it processes the food on the next time step. Starting with an empty gut, the limpet gains no energy until the time step *after* it starts feeding, which is beyond our one-step time horizon. Therefore, feeding never improves expected fitness and energy reserves fall until they reach zero.

How can we deal with this complication that feeding prior to the current time step drives energy assimilation? The solution involves prediction: making our model limpet aware that food in its gut will turn into future energy. The simplest way to do this uses a time horizon longer than one step, so food consumed on the current time step turns into energy before the time horizon.

We now change our formulation so that a limpet selects the activity that provides the highest expected energy reserves at a sliding time horizon two time steps (1 hour) in the future. It uses the simplest possible prediction: that activity and conditions (the value of *favorable?*) in the second time step remain the same as in the current time step.

This formulation produces the same basic dynamics as the DSVM model: the model limpet starts feeding at the beginning of each day, alternates between feeding and stasis through the favorable period, and rests during the unfavorable

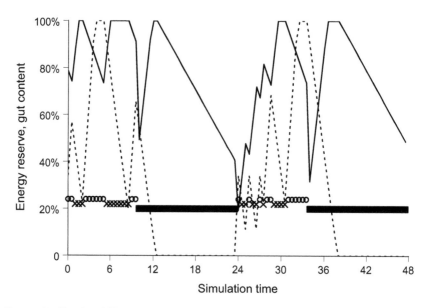

FIGURE 6.1. Results of SPT version 1 of the limpet foraging model, with 10 hours of favorable conditions per day. The graph shows the first two simulated days; behavior on remaining days repeats that shown here for hours 24–48. The lines indicate the limpet's energy reserves (solid) and gut content (dashed) at the end of the time step, as a percentage of the maximum. The symbols indicate the activity chosen on each ½-hour time step: open circle = feeding; × = stasis; and solid black bar = resting.

period (figure 6.1). (The switch to resting during the unfavorable period is of course strongly imposed by the assumption that limpets die if they do anything else.) During the favorable period, the limpet feeds long enough to fill its gut, then switches to stasis until its gut is again nearly empty.

How well does this model reproduce the observed patterns that it was designed to explain (section 6.3)? A simulation experiment varying the length of the favorable period from 3 to 20 hours produced exactly 1 hour of activity (feeding or stasis) per hour of the favorable period: the model limpet was always active in the favorable period and never active outside it. This result reflects the most general trend in the observed patterns but not their details. Further, this formulation does not cause the limpet to maintain the highest possible energy reserves through the unfavorable period by filling its gut at the end of the favorable period. These deviations from what seem like optimal behavior appear to result from the short time horizon we use: "knowing" that (1) filling the gut at the end of the favorable period produces higher reserves through the rest of the day or (2) being active throughout long favorable periods is not valuable requires basing decisions on predicted consequences over longer time horizons. Let us try that next.

6.5 SPT VERSION 2: MAXIMIZING MEAN EXPECTED ENERGY RESERVES UNTIL DAY'S END

If a 1-hour time horizon seems too short, it seems reasonable to try the end of the current day instead. The end of the day is a natural time horizon under the assumptions we took from Santini et al., because tidal conditions switch from unfavorable to favorable at the beginning of each day. We no longer have a "sliding" time horizon that is always the same length of time in the future but instead a "leaping" horizon that is always the next midnight.

As we lengthen the time horizon, let us also now use the mean energy reserves throughout the day as our fitness measure. This addresses the arbitrariness inherent in the limpet model when using the energy reserves at the end of the day as the fitness measure; many different combinations of behavior during a day can yield the same final energy reserves. Our fitness measure is now $S_T \overline{w_T}$, where $\overline{w_T}$ is predicted mean energy reserves between the current time step and T, midnight of the current day.

We must now determine how the limpet should predict its activity, energy reserves, and survival during future time steps of the current day. Assuming (as we have in almost all previous example models) that the same behavior will be used until the time horizon will not succeed: the limpet would starve if it fed every time step (because it loses energy when feeding) or if it used stasis each step (because it would never obtain food). Instead, we can have the limpet predict future behavior by assuming it alternates between feeding and stasis on all remaining time steps in the favorable period (if any) and rests on all remaining unfavorable time steps. This prediction will cause the limpet to feed only when necessary: if it can use resting or stasis and still maximize expected energy reserves over the rest of the day, it will.

The activity submodel of this second SPT formulation therefore uses the following steps:

- Predict the energy reserve and gut content on each remaining time step of the day if it feeds on the current time step, alternates between feeding and stasis on any remaining steps of the favorable period, and rests on all remaining steps of the unfavorable period. The prediction assumes activity switches to stasis whenever continuing to feed would exceed the gut's capacity, and switches back to feeding if the gut has room for the intake. These calculations are done by iteratively applying the energetics submodel for each remaining time step. (Iteration is necessary to enforce the maximum limits on gut contents, gut processing rate, and energy reserves at each predicted time step; these limits strongly affect results.)
- During the iterative prediction of energy reserves for feeding, also predict the probability of surviving until the end of the day. Predicted survival is set to zero if predicted energy reserves fall to zero at any point, and is otherwise one.
- Calculate the fitness measure for feeding activity as predicted survival to day's end times predicted mean energy reserves between the current time and day's end.
- Calculate the fitness measure for stasis by repeating the above steps, but this time assume the limpet uses stasis on the current time step, alternates between feeding and stasis on the remaining favorable time steps, and rests on the remaining unfavorable steps.
- Calculate the fitness measure for resting by repeating the first three steps while assuming the limpet rests on the current time step, alternates between feeding and stasis on the remaining favorable time steps, and rests on the remaining unfavorable steps.
- Select the activity that provides the highest value of the fitness measure, while still using the rules from version 1 for what to do if more than one activity provides the same fitness.

This formulation produces behavior that reproduces the basic observed patterns: the limpet alternates between feeding and stasis throughout the favorable period, and always fills its gut just before the unfavorable period starts (figure 6.2). Energy reserves dip during the bouts of feeding necessary to refill the limpet's gut and at the start of the unfavorable period, as a result of the cost of moving to the resting site.

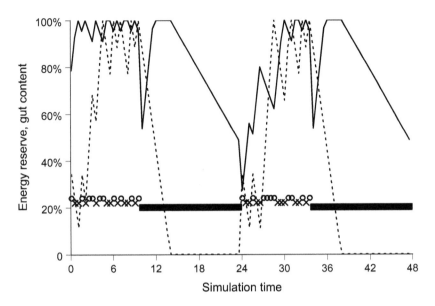

FIGURE 6.2. Simulated limpet behavior for the first two days using the second version of SPT, with 10 favorable hours per day. Format as in figure 6.1.

Now, let us see whether our SPT model can reproduce the two more complex patterns observed in real limpets: foraging during the unfavorable period when the number of favorable hours T_{PAP} is low, and not foraging during all favorable hours when T_{PAP} is high. Santini et al. (2014) reproduced these patterns with their DSVM model by assuming that mortality is less than 100% if a limpet is active during the unfavorable period. With the same assumption, our SPT model at least coarsely reproduces these patterns (figure 6.3). When T_{PAP} exceeds 20 h, whether the limpet chooses feeding or resting is unimportant because the unfavorable period is too short for a resting limpet to empty its gut. When T_{PAP} is very short, relatively low risks of feeding during unfavorable conditions can be offset by the energy gain. (We expect that adding more realism to risk and energetics would allow the model to reproduce these patterns more robustly.)

6.6 CONCLUSIONS

This example again illustrates that SPT can produce behavior closely resembling that of DSVM, and—importantly—that its freedom from some of DSVM's restrictive assumptions can actually let SPT produce behavior that seems more realistic and conveys higher fitness. In the limpet model, our "leaping" time horizon in

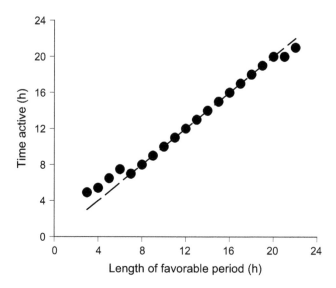

FIGURE 6.3. Predicted active hours per day for favorable period lengths of 3 to 20 hours, with survival probability per time step of 0.95 when foraging in the unfavorable period. Circles represent model results and the dashed line represents 1:1 correspondence between time active and favorable period length. To make it possible for a model limpet to survive at favorable periods < 8 h, g_{max} was increased to 300 mm^3 and δ was increased to 2.0 J/mm^3, but no other calibration was done.

place of DSVM's fixed time horizon resulted in less arbitrary behavior and provided higher fitness in the sense that the SPT limpet kept its energy reserves high throughout the simulation, not just at the end.

Nonzero survival during the unfavorable period was trivial to add to the SPT model and necessary for reproducing patterns observed at very low values of T_{PAP}. However, a longer time horizon should be used with nonbinary survival. (As we discuss in section 8.3, any creature that makes trade-offs against survival one day at a time is unlikely to live long.) A fitness measure that retains the benefits of a "leaping" daily time horizon but considers survival over N future days can assume that the predicted mean energy reserves for the current day w_t and predicted survival for the current day S_t would also be obtained on future days; this fitness measure would therefore be $w_t S_t^N$.

The simplifications in the limpet model of Santini et al. (2014), such as binary survival and maximum values of gut contents and energy reserves, limit the complexity of its results. These simplifications were probably motivated in part by the need to make the DSVM optimization easier to formulate and solve. Using SPT reduces the computational tractability problem and allows the addition of "details" that often are important for reproducing observed patterns. Potential complexity that could readily be added to our limpet model includes explicit representation of space, spatial variation in food quantity or quality, competition among limpets, and the effects of additional energy demand for reproduction on foraging behavior.

We found the limpet model of Santini et al. (2014) when searching for applications of DSVM that we could convert into examples of SPT. It turned out to be unexpectedly interesting because slow food processing causes a time lag between the execution of a behavior and some of its consequences for fitness. Other examples of such time lags include plants committing to growth of new structures and any organism committing to production of reproductive organs. Modeling such behaviors as adaptive trade-offs requires explicit prediction of the future benefits and costs of the decision alternatives.

CHAPTER 7

Example Four: Facultative Anadromy in Salmonid Fishes

7.1 INTRODUCTION AND OBJECTIVES

In this chapter we present our fourth and final example model, with the objectives of (1) illustrating the application of SPT to a new kind of decision—a life history decision—in a case where DSVM has been applied successfully; and (2) describing the unique ability of models utilizing SPT to address population-level questions of particular interest to conservationists and managers. In this case, SPT produced individual-level decisions similar to those of DSVM, but including them in a population-level model led to quite different conclusions than those implied by the individual-level DSVM analysis.

Salmonid fishes exhibit amazing life history diversity. For example, Hodge et al. (2016) identified 38 different life histories in *Oncorhynchus mykiss* in the Klamath River of northwestern California. One fundamental distinction among salmonid life histories is whether or not individuals migrate to the ocean: *O. mykiss* that do so are called steelhead and those that remain resident in freshwater are called rainbow trout. In some species the anadromy versus freshwater residency decision is fixed, and the decision appears to have a genetic component across salmonid species. However, within some species, including *O. mykiss*, the decision is not under strict genetic control: for example, resident fish commonly produce anadromous offspring (Courter et al. 2013). In general, facultative anadromy can be seen as an adaptive behavior that trades off the fitness benefits of going to the ocean (larger size and fecundity) versus those of remaining resident (shorter time to reproduction and avoidance of the risks of migration) (Gross et al. 1988).

The anadromy versus residency decision is important to fish conservation and resource management. At the population level, the retention of life history diversity can contribute to sustainability via *portfolio effects* (Schindler et al. 2010). The choice between anadromy and residency can also affect whole ecosystems, because where and how much the fish feed has effects on food webs in both the ocean and freshwater (Wipfli and Baxter 2010).

In the specific case of *O. mykiss*, anadromous individuals can be particularly highly valued such that people responsible for conservation and management seek strategies to promote that life history. This conservation and management goal highlights a key contrast identified in this monograph: the individual level of adaptive decision-making versus resulting population-level outcomes. For example, in this case promotion of the decision to become anadromous at the individual level need not result in the desired population-level outcome of more anadromous fish: if lower freshwater survival increases the proportion of individuals choosing anadromy, *higher* freshwater survival might actually produce more anadromous fish because more of the individuals choosing anadromy survive to reach the ocean.

Facultative anadromy has been the subject of much empirical research and modeling. A valuable conceptual framework has been developed to describe it as an adaptive behavior of individuals (e.g., Mangel and Satterthwaite 2008). This framework assumes that the anadromy decision (1) maximizes reproductive success through effects on survival and size at spawning, (2) is influenced by an individual's current size and growth rate, (3) is made in advance of its implementation (fish decide to migrate to the ocean long before they actually leave their natal stream), and (4) affects behavior during the time between decision and implementation.

In sections 7.2 and 7.3 we first describe the DSVM model that provides the conceptual framework for the anadromy behavior, then summarize how we incorporated that framework in an inSTREAM-based population model by using SPT. Section 7.4 then discusses how these two implementations of the same conceptual framework produced similar individual adaptive behavior but sometimes different conclusions about population responses to management.

Because of this model's complexity, we do not describe it completely in ODD format here or provide its computer implementation. A complete description of the model is available as supplementary material to Railsback et al. (2014) and from this book's web site as described in the preface.

7.2 THE DSVM MODEL

Satterthwaite et al. (2009, 2010) modeled facultative anadromy in *O. mykiss* from the perspective of optimal individual decision-making in a DSVM framework. The DSVM approach applied the conceptual framework described above, and further assumed that fish make life history decisions within specific time windows. For example, the decision on migration to the ocean is made at the end of a two-month window, three months in advance of when outmigration would actually occur.

The model identifies optimal life history decisions from the size of fish at the end of the window and its growth during the window. Using empirical estimates of stage-specific mean growth and survival, the DSVM model predictions of optimal life history alternatives corresponded well with patterns of observed life histories in both coastal streams and Central Valley rivers in California.

7.3 THE IBM USING SPT

We were motivated to build an individual-based, spatially explicit, population-level model incorporating facultative anadromy for several reasons. First, population-level outcomes are of primary interest to fisheries managers, so a model that directly yields population-level results should be useful. Second, anadromy decisions by individuals are subject to feedbacks: they may depend on their interactions with other individuals. For example, in a situation where many individuals decide to migrate to the ocean, lower competition among those who remain may affect their growth, survival, and therefore their anadromy decisions. Third, including variation in the physical environment could be important for capturing the consequences of individual decision-making at the population level because, even within stream reaches, growth opportunities and mortality risks can vary dramatically over space and time. In the IBM, the survival probabilities and growth rates that drive the anadromy decision are not assumed as parameters but instead are outcomes of habitat characteristics, behavior, and interaction among individuals. Finally, this kind of model has the potential to estimate the consequences of environmental change for fish growth and mortality risks, setting the stage for exploration of how such change might affect populations of facultatively anadromous individuals. For all these reasons, we believed an IBM could effectively address some critical management questions: Does enhancement of stream habitat increase or decrease production of anadromous steelhead? Could making freshwater habitat safer and more productive tip the fitness balance toward the resident life history and reduce numbers of anadromous individuals?

The SPT model of facultative anadromy by Railsback et al. (2014) extended the inSTREAM family of salmonid IBMs that use SPT to model habitat selection (Railsback et al. 1999; Railsback and Harvey 2002) as described in chapter 2. In these models, model trout select among tens to hundreds of habitat cells, with their decision depending on multiple state variables (reproductive status, length, weight) as well as dynamic environmental variables (energy intake and predation risk in each cell depend on variables of the individuals and habitat cells: on daily flow, temperature, and turbidity and on the number and size of competing trout). Fish choose habitat daily to maximize their anticipated fitness under the assumption that current conditions will continue to apply. The SPT model of facultative anadromy adds to this framework a contingent, irreversible life history decision, determined by the growth and survival probabilities experienced by individual fish. The SPT model relies on the conceptual model of individual decision-making applied in DSVM by Satterthwaite et al. (2009, 2010), which we sought to closely follow wherever we could.

One component of the DSVM approach we did not seek to match was incorporation of many *O. mykiss* life history alternatives generated by the fact that fish can commit to anadromy or residency over a range of ages. Instead, the SPT model of facultative anadromy limits the life history alternatives for *O. mykiss* to the two most common ones: to remain resident intending to spawn at age two or to migrate to the ocean within the first year after birth. During the "stay or go" phase of life, each virtual juvenile fish still in its natal stream decides, every day, whether to remain a resident or make the irreversible decision to become anadromous. A fish decides to become anadromous if its fitness measure (expected offspring production at next spawning) is higher for anadromy than for residency. To evaluate these fitness measures, the simulated fish must predict their size at future possible migration and spawning dates, and their probability of surviving until those dates. Individual growth and survival emerge from complex interactions among fish and their highly variable habitat.

Accumulating differences among individuals and potentially strong temporal variation in habitat conditions make undesirable the assumption that growth and survival rates are the same for all fish and all days. Instead, individual fish use their own recently experienced growth rates and survival probabilities to predict future size and survival. From the time it becomes a free-swimming individual, each fish maintains a memory of growth and survival probability for a set length of time—we use the past 30 days, which limits the influence of short-term events but also is short enough to capture seasonal variation. These records of growth and survival probability are used in the daily computations of fitness that determine whether a fish remains resident or becomes anadromous.

Individuals first have the option of migrating to the ocean at an age equal to the length of their memory of growth and survival probability. They retain the option of becoming anadromous until the date by which the commitment to spawning as a resident must be made. This date is determined by the first possible date of spawning and a "maturity decision interval" that reflects the time needed for the physiological changes required for spawning. The model also incorporates an interval between the decision to become anadromous and outmigration to the ocean (the "smolt delay" parameter). These delays between the commitment to life history events and the events themselves reflect the conceptual framework for facultative anadromy also included in the DSVM models.

   The fitness measures for residency and anadromy estimate the expected numbers of future offspring: the product of the probability of survival to reproduction and expected fecundity. Resident fish estimate their probability of survival to
reproduction using the averages of their daily survival probabilities in memory,
for starvation and nonstarvation risk, each taken to a power equal to the number of
days until spawning at age two. Fecundity is determined from a relationship with
spawner length. The length of the fish at spawning is projected from its current
length and the mean of the growth rate over the fish's memory. For anadromous
fish, survival until outmigration is computed similarly to the survival computations
for resident fish, except that the exponent is the value of the parameter describing
the number of days between when a fish decides to outmigrate and when it actually does. The computation of ocean survival is modeled as a logistic function of
fish length at outmigration following Satterthwaite et al. (2010), with the length
determined by current length and the mean growth rate in memory. (Survival of
outmigrants generally improves with fish size [e.g., Thompson and Beauchamp
2014].) Because outmigrant size does not clearly predict future fecundity, the
expected number of offspring for anadromous fish that successfully reproduce is
simply assumed constant, with separate values for females and males.
   Once each simulated fish makes its anadromy decision by using SPT and its
survival and growth experience resulting from past habitat selection, that decision in turn affects future habitat selection. Fish destined to remain resident in
freshwater maximize a fitness measure equivalent to the one used to evaluate the
resident alternative for the life history decision (expected survival to spawning
at age two times expected fecundity), except that for habitat selection fish use
daily cell-specific values of growth and survival rather than utilizing memory
of their recent experience. Fish destined to migrate to the ocean use a fitness
measure similar to the one they used to evaluate the anadromous life history alternative (expected survival through the ocean life stage times anadromous fecundity), except that the time horizon is the number of days until migration begins
and again the habitat selection decision compares cells by their current growth
and survival conditions. The fitness measure used by migrants gives them strong
motivation to grow because of the positive effect of fish length on ocean survival.
Growth acceleration before migration has been observed in real salmonid fishes
(Metcalfe et al. 1998).
   We conducted simulations that included natural variation in physical conditions on multiple spatial and temporal scales. Simulations covered 17 linked
sites on a major tributary of the Sacramento River in Northern California that
totaled about 5 km of stream represented by over 13,000 habitat cells. Simulations spanned 6 years that included a wide range of spawner densities. Baseline simulations used values of environmental parameters with critical effects

on fitness—food availability and predation risk—calibrated using site-specific information on fish abundance and growth.

7.4 SPT MODEL RESULTS AND APPLICATIONS

At the level of individual fish, we can compare facultative anadromy decisions produced by SPT to the "optimal" decisions produced by the DSVM model. However, the SPT model, unlike DSVM, also produces population-level predictions that in this case could lead to different management conclusions than the individual-level results. At the individual level, the SPT model yielded patterns of facultative anadromy that paralleled the results of the DSVM optimization approach of Satterthwaite et al. (2009, 2010). In both models, larger juveniles chose anadromy given any positive growth rate. Smaller fish tended to remain residents except for those benefiting from high growth rates. And in both models, at least under some conditions, higher survival in freshwater made fish less likely to migrate to the ocean.

The IBM, therefore, produced individual adaptive behavior similar to that of DSVM but in a framework that included interactions among variable individuals under realistically dynamic habitat conditions. Consequently, the IBM produced much more complex and variable results. Both resident and anadromous individuals occurred over a wide variety of conditions. Even over broad ranges of food availability and freshwater survival, there were always some fish that chose anadromy, instead of anadromy occurring only when food availability was high or survival low. This kind of result makes sense with recognition of the importance of individual and habitat variability. Even under generally poor conditions, some individuals may find and predominate in habitat patches allowing for relatively high growth likely to lead to anadromy. And under high-growth conditions that generally favor anadromy, late-emerging individuals can face substantial competition that could make their best decision to select patches offering low growth but also low predation risk. Such fish would probably become residents. Population-level outcomes of the IBM included high numbers of both anadromous and resident fish under some conditions (high food availability and moderately high survival) and low numbers of both in others (low freshwater survival).

The IBM also allows forecasting of the population-level effects of environmental change. For example, given that the adaptive behavior makes anadromy less favorable as freshwater survival increases, will restoring freshwater habitat to improve survival result in fewer highly valued anadromous fish? For the *O. mykiss* we simulated, the IBM predicted that even as a lower percentage of individuals chose anadromy under conditions providing higher survival, more of those choosing anadromy lived long enough to emigrate, so anadromous abundance actually increased. Narrow focus on individual decision-making has led managers to consider degrading freshwater conditions to promote anadromy; tracking population-level outcomes using models of interacting individuals indicates that a more reasonable approach to promoting anadromy would be to provide more favorable conditions for growth—an approach that would also produce more residents. By including individual adaptive behavior in a population-level model, the IBM that includes facultative anadromy allows prediction of responses in real-world settings that people really care about, in this case the abundance of anadromous fish.

CHAPTER 8

Guidance for Using State- and Prediction-Based Theory

8.1 INTRODUCTION AND OBJECTIVES

In the previous chapters we provided an introduction to SPT and example applications. Now we attempt to distill what we have learned from our experience into guidance on using SPT to build models of populations and communities of adaptive individuals. This chapter therefore includes a section for each of the five steps for using SPT outlined in chapter 3, providing as much guidance as we can on the steps unique to SPT. We give less attention to steps that use techniques and theory from DSVM, life history theory, or related fields of theoretical and behavioral ecology and those already thoroughly covered in the individual-based modeling literature.

The most important aspect of SPT to remember is that we are not trying to build optimal, or even necessarily accurate, models of how an organism's behavior affects its future fitness. Instead, we are trying to find simplistic models that produce *realistic behavior* in contexts where optimization is impossible. This means that an SPT model of how an individual estimates its future fitness for decision alternatives uses simple predictions that we know are usually wrong, and approximations that we know are inaccurate. We do this in part to keep our models computationally feasible when modeling populations of individuals that each make frequent complex decisions. But the main reason we accept these simplifications is that in realistic settings—populations of interacting and competing individuals, in variable environments, possibly responding to predators or prey that also adapt—there is no such thing as optimal behavior. Instead, modelers using SPT work within the same constraints that the modeled organisms contend with, and use reasonable assumptions about what information individuals sense and how they use it in unpredictable conditions. We should also keep in mind, though, that we learned in chapters 4 and 5 that SPT, including major simplifications, can produce near-optimal behavior.

While SPT can be used like DSVM, as a framework for thinking about and modeling how an individual makes a particular decision, its main purpose is to model adaptive trade-off decisions in individual-based population models; this is a book on population modeling, after all. Therefore, we assume here that using SPT is part of the larger process of developing, analyzing, and applying an IBM to address population-level questions, and our five steps therefore include that process.

Please be aware that these five steps need not be conducted in sequence. Instead, modelers will undoubtedly think about all of them as they proceed. Keep in mind as well that our experience is limited and much remains for ambitious researchers to discover about using this approach!

8.2 STEP 1: DEFINING THE DECISION THAT SPT MODELS

Productive modelers always start a new model by carefully defining the problem it is intended to solve. For SPT, step 1 is therefore to define exactly what adaptive trade-off decision the model individuals make: What alternatives do they choose among, what variables differentiate those alternatives, and what elements of individual fitness are affected by those variables? We answer these questions by designing the IBM that SPT will work within.

We and others have published extensive guidance on developing and analyzing IBMs, especially Grimm and Railsback (2005) and Railsback and Grimm (2019). The main strategy we advocate for model development and analysis is *pattern-oriented modeling* (POM)—the use of a variety of patterns observed in real systems to (a) determine what should and should not be in the model, (b) test and improve the theory for adaptive behavior, and (c) estimate parameter values. Grimm et al. (2005), Grimm and Railsback (2012), and Railsback and Grimm (2019) discuss POM thoroughly, and Railsback and Johnson (2011, 2014) provide a detailed example application. In step 1 we use observed patterns to guide model design, using them as filters for determining what should and should not be in the IBM. In step 5 we use the patterns to test and refine our SPT-based behavior submodel.

Here we briefly outline IBM development phases that precede and are essential to developing the SPT theory for adaptive behavior. We return to the whole IBM as part of the last step of SPT development, in section 8.6. The IBM development process outlined here is based on extensive experience by many people, and designed to be efficient and to avoid common pitfalls. While we have seen many projects rapidly develop useful IBMs, we have also seen IBM projects fail, or take much longer than they should have, because modelers jumped into the details of programming a too-complex model without first working their way through the initial phases to figure out what should and should not be in the model.

State the problem that the IBM addresses. The lack of a sufficiently clear and specific statement of the problem to be modeled is almost certainly the most common reason modeling projects take much longer than they should, or fail completely. State exactly what management question or ecological problem the model is intended to address, in what systems, and what related problems it is *not* intended to address. The greater the clarity and specificity of this statement, the more useful it is in building the model, mostly by helping the modeler identify details and complexities that can and should be omitted. Whenever decisions about model design seem difficult or ambiguous, go back to the problem statement and try to make it clearer and more specific.

In our experience it is much more productive to model specific, real systems (as the trout and *Daphnia* models did) instead of "general" or hypothetical ones, such as the forager model of chapter 4. Modeling a real system allows the use of observations and data (including qualitative patterns, as explained below) throughout model design, including the choices of submodels and parameter values. Working on a real system also makes it harder to avoid the key behaviors and complexities that actually drive ecological systems, raising the likelihood of producing innovative and meaningful results.

Identify key population drivers, and patterns that characterize the system with respect to the problem being modeled. This phase uses experience, literature, and judgment to identify external drivers (habitat and resource availability, predation risk, harvest, etc.) and individual-level behaviors and mechanisms (habitat selection, energy allocation, etc.) that seem important for solving the modeling problem, and to identify the observed patterns to use in POM. These patterns characterize how the system and its individuals respond to the drivers and mechanisms that we believe are essential for solving the modeling problem, and are usually obtained from both field studies and literature. (We discuss selection of patterns further in section 9.2.) At this phase, useful patterns are typically qualitative trends in how either the population or its individuals respond to changes in the external drivers. To help identify useful patterns, think of them as criteria for determining the readiness of the model to address the problem it is designed for (or as a "stopping rule" for when model development and testing can stop and analysis phases such as calibration and validation can start): if the model does not reproduce these patterns, it is not ready for solving the problem.

Design the model's structure and schedule, and identify essential adaptive behaviors. This phase establishes the "overview" part of the ODD model description, including identification of appropriate spatial and temporal scales, necessary entities and the state variables they must have, and what processes and individual behaviors change the state variables. The model's schedule should also be determined; it specifies which entities execute which actions, in what order.

POM is very important at this phase; it uses the model's purpose and the characteristic patterns identified previously as criteria for making these critical model design decisions in such a way that the model contains just enough complexity to serve its purpose. Even for modelers not intending to use POM explicitly (e.g., because they believe there are not sufficient observed patterns), we strongly recommend absorbing the pattern-oriented model design literature cited above (especially the example of Railsback and Johnson [2011]) to learn how to make model design decisions thoughtfully; intuition and familiarity are not always good guides. The "design concepts" element of the ODD model description protocol we use in chapters 4 through 6 (Grimm et al. 2010; Railsback and Grimm 2019) is also very helpful as a framework for thinking about essential questions in designing IBMs.

This phase identifies the specific behavior to be modeled using SPT. Usually, an effective IBM includes a very small number of adaptive individual behaviors. As the number of such behaviors increases, the complexity of the model and the modeling process (time and effort to design and test the behavior models and to calibrate, analyze, and understand model results) increases rapidly. Our favorite models have only one adaptive trade-off. Deciding that more than one trade-off behavior is essential to a model can be a sign that the model's purpose needs to be further constrained, at least until a first version is fully analyzed. (For example, the facultative anadromy decision of chapter 7 was added to our salmonid models only after their primary adaptive trade-off behavior—habitat selection—was thoroughly tested.)

Identifying the adaptive trade-off behavior to model means determining what decision the individuals make, what alternatives they consider, and what variables—external and internal—affect the decision. Completion of this phase includes at least tentatively defining, for each adaptive trade-off behavior to be modeled:

- Which individuals execute the behavior, in what order, and when, in the model's schedule of actions;
- The alternatives that the individuals evaluate;
- The fitness elements—predation risk, starvation risk, reproductive output—that drive the decision; and
- The variables of both the decision alternatives and individuals that determine those fitness elements.

When these model characteristics have been defined, the adaptive behavior is defined well enough to start designing the SPT theory.

8.3 STEP 2: SELECTING FITNESS MEASURES
AND TIME HORIZONS

The second step in applying SPT is to develop a fitness measure, which includes selecting an appropriate time horizon. The fitness measure is a simplified measure of the relative fitness that decision alternatives provide to the individual at a future time horizon. The fitness measure should include only those elements of fitness relevant to the decision being modeled. By "elements," we refer to objectives an individual must achieve to be reproductively successful, especially survival of predation, survival of starvation, and either reproductive output or accumulation of reproductive resources, such as adequate size, energy reserves, or mates.

Developing fitness measures for SPT is essentially the same problem as developing them for DSVM, except for the time horizon issue addressed below. The extensive DSVM literature is relevant and we do not attempt to replace it here. As a reminder, this book has presented these fitness measures:

- The trout model's habitat selection behavior (chapter 2) uses the expected probability of surviving both predation and starvation until the time horizon. For juveniles, the fitness measure includes a term representing attainment of reproductive size: the fraction of minimum spawning size attained by the time horizon.
- The forager model (chapter 4) uses the probability of surviving both predation and starvation until the time horizon.
- The *Daphnia* model (chapter 5) uses expected reproductive output at the time horizon, evaluated as the probability of survival until the time horizon multiplied by the expected reproductive output attained by the time horizon. The first version of this model represented reproductive output as accumulated mass allocated to reproduction, while later versions represented output as the number of offspring produced.
- Activity selection by limpets (section 6.3) uses expected energy reserves over the remainder of the current day, evaluated as the mean of predicted reserves for each time step remaining in the day multiplied by the expected probability of surviving until the end of the day.
- The *O. mykiss* life history decision (chapter 7) uses expected reproductive output at the next spawning season, evaluated as the expected number of eggs produced (or fertilized, for males) times the probability of surviving both predation and starvation until spawning.

Probabilistic nature of fitness measures. The fitness measures we have used are all probabilistic: they are either a probability of surviving until the time horizon, or a survival probability multiplied by something else (like number of offspring). Probabilities have several advantages as fitness measures compared to alternatives such as integers (e.g., the number of offspring) or Boolean values (e.g., an individual either does or does not survive until the time horizon). We discussed one advantage in section 4.5: expressing fitness measures as probabilities allows evaluation of alternatives even when none are good. This ability is essential in an IBM, because populations often encounter stressful periods when most individuals have no good options and only the lucky survive, but the population persists. Successful adaptive behavior is extremely important in these situations—in models as well as in reality. If we assume a fitness measure has a value of zero anytime energy reserves reach zero (using the example from chapter 4), then model individuals will be unable to distinguish between alternatives that provide rapid vs. slow starvation and, therefore, low vs. moderate hope for survival. If instead we define the starvation component of the fitness measure as a survival probability that decreases continuously as energy reserves decrease and weight loss accumulates, we give the individual the ability to identify and choose alternatives that provide higher probability of surviving until conditions improve. (And conditions can improve rapidly and unexpectedly; e.g., an individual's situation can improve instantly if its competitors die.)

A second advantage of basing fitness measures on probabilities is the potential to use probability theory to evaluate them. For example, we routinely use logistic functions to represent the probability of surviving various kinds of risk. Such functions can be based on field observations of discrete mortality events and logistic regression. We further discuss use of probability theory in SPT in section 8.4.

Fitness measures for life history decisions. Life history decisions can include whether or when to advance to the next life stage, when to start reproduction, or (as in the *O. mykiss* model of chapter 7) selecting between alternative life history strategies. Several characteristics of life history decisions set them apart from the routine foraging decisions that we otherwise focus on: they may be made only once or occasionally, are irreversible, and compare alternatives that affect the individuals in very different ways. Different life history alternatives may expose individuals to different predation risks and food resources; some alternatives may require expensive transitions, such as physiological transformation and migration; and alternatives may differ in the timing and amount of reproductive output.

To compare life history alternatives that differ in these ways, we need a fitness measure that can be evaluated for each alternative despite their differences. The anadromy decision in our *O. mykiss* model compares the alternatives of migration to the ocean and remaining there for several years before returning to reproduce versus staying in a river and reproducing the next year. In the model, once an individual selects one life history it cannot switch to the other. To compare these two alternatives, we are therefore restricted to a fitness measure evaluated at the time of reproduction. Hence the fitness measure is the expected number of eggs at next reproduction: the product of predicted survival until next reproduction and predicted fecundity. This measure ignores complexities such as how the time until spawning and the potential for repeat spawning affect lifetime reproductive output. We then use separate, and highly simplified, methods to predict those fitness elements. This approach of using an "ultimate" fitness measure—based on survival to and fecundity at a specific time of future reproduction—is likely to be a useful approach for evaluating divergent life histories.

Time horizons. Selection of an appropriate time horizon is obviously critical to a successful SPT fitness measure and is a more open-ended issue for SPT than for DSVM. Because DSVM uses a one-time optimization of behavior over the entire time up to the time horizon, it requires specifying the time horizon as a fixed future time. SPT, however, repeats and updates decisions at time steps much shorter than the time horizon, so the time horizon can change over simulated time. The simplest example of a changing time horizon is the "sliding" time horizon used in the trout and *Daphnia* models: on each time step, the individual makes decisions by evaluating a fitness measure that is a constant number of time steps (90 days; 240 hours) into the future. (We call this fixed number of time steps the "length" of the sliding time horizon.) Sliding time horizons avoid the potential problem of "terminal effects"—changes in behavior that result from how the relative importance of predation vs. starvation changes as the time horizon is approached (section 5.3). Sliding time horizons are therefore appropriate for individuals not approaching a life history event or other transition that changes how they make decisions.

Selection of the length of a sliding time horizon is a critical decision that should consider several issues. One basic consideration is that trade-offs involving mortality risks, including starvation, require time horizons longer than the time step of most models. A key lesson from DSVM is that good trade-off decisions cannot be based only on the probability of surviving the current day (hour, etc.): what might be a safe bet once (e.g., accepting a 1% risk of mortality in exchange for a good meal) can yield a very poor plan for the future (1% risk per time step equals an expected life span of < 70 time steps). Behavior theory that considers only current risks, such as the "minimize μ/G rule" discussed in section 1.3, is unlikely to give adequate weight to survival.

Another major consideration is how time horizon length depends on the methods used to predict conditions until the time horizon (addressed in section 8.4). For example, the prediction method we recommend trying first—simply assuming that current conditions will persist until the time horizon—makes behavior more responsive to current conditions when combined with a long time horizon: small differences among alternatives in survival have strong effects over long time horizons. But a prediction method that assumes conditions follow a long-term trend or always revert toward "normal" may produce behavior *less* responsive to current conditions with a long time horizon compared to a shorter one.

Our trout model provides a second example of how time horizon length and prediction method interact to affect the relative importance of different fitness elements. Fish select habitat to trade off avoidance of short-term risks to survival, such as predation and the need to obtain food. The prediction method for predation assumes that the current survival probability persists until the time horizon. However, for starvation the model trout assumes that the current rate of weight gain or loss persists until the time horizon. Under these assumptions, starvation is unlikely over short times, even when fish are underweight; but as the time horizon gets further away, the predicted probability of an underfed fish starving before reaching it increases.

To reflect the behavior of real organisms under most conditions, the influence of both short-term risks and the risk of starvation indicate a reasonable time horizon length. In general, when decisions involve trade-offs between immediate risks, such as predation, and longer-term risks, such as starvation, the time it takes an initially healthy but unfed individual to starve can be a reasonable starting point for a sliding time horizon. Another consideration in determining time horizon length is the trade-off between the accuracy of prediction over the time horizon and the relationship of the fitness measure to total "lifetime" fitness. The longer the time horizon, the less accurate the individual's predictions of its fitness measure are likely to be, which may or may not be important to producing realistic adaptive behavior. Longer time horizons may better represent lifetime fitness, especially for species that reproduce rapidly, but again this consideration may or may not be important for producing realistic behavior.

Population-level patterns can also inform selection of fitness horizon length. For example, virtual animals exhibiting reasonable fitness-seeking behavior should persist under conditions that allow persistence of natural populations. Unreasonable time horizons can be revealed by frequent extinction of populations in simulations of persistent natural populations. We also have observed that, in simulations including realistic spatiotemporal variation in environmental conditions, time horizon length influences the variation in reproductive success among virtual individuals that could be thought of as "genetically" identical (they all use the same algorithms). In our trout model, reasonable time horizons produce strong variation in reproductive success among individuals observed in natural populations (e.g., Kanno et al. 2011), while shorter time horizons yield individuals too risk averse to generate this variability.

The "leaping" time horizon used in the limpet model (chapter 6) illustrates a way for the time horizon to match a cycle in environmental conditions. When such cycles strongly regulate individual behavior, the cycle end makes a natural time horizon. However, modeling trade-offs involving survival often requires converting expected fitness over one individual cycle into fitness over a longer time horizon, as discussed in section 6.6.

Fixed time horizons can be appropriate in some applications of SPT. The *O. mykiss* life history decision uses as its time horizon the fixed date that marks the start of the next spawning season. Therefore, the individuals' behavior changes as that time horizon approaches, which reflects that, for real fish, spawning is a major event that requires preparation and recovery.

With SPT, we can even make the time horizon an outcome of decisions as well as part of the decision model. Consider a model in which individuals make frequent foraging decisions that trade off predation risk and growth, and then reproduce when they achieve a certain size. The foraging decision could use a very simple fitness measure—survival of predation until reproductive size is achieved. The time horizon is therefore the date on which the individual expects to reach reproductive size: the more it eats, the sooner it reaches reproductive size and therefore the shorter the time horizon. The shorter the time horizon, the higher the probability of surviving to it (figure 8.1). (This very simple approach has

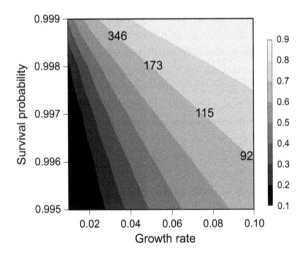

FIGURE 8.1. Value of a fitness measure that represents expected survival until attaining a size $s = 10$, for an individual currently with size $s = 1$. The X axis rate is growth in size units per time step, and the Y axis is survival probability per time step. The time horizon is therefore $T = (10 - s)/(\text{growth rate})$ and the fitness measure's value is (survival probability)^T. The labels within the plot indicate the time horizon at four points that all provide fitness of 0.7.

limitations if an individual might temporarily be forced to choose from alternatives that all offer zero or negative growth. The fitness measure is undefined when growth is zero or negative, so the individual would not be able to identify the least bad among the negative-growth alternatives.)

8.4 STEP 3: MODELING PREDICTION OF ENVIRONMENTAL CONDITIONS AND FITNESS ELEMENTS

The third step of SPT designs the assumptions and specific methods that model individuals use to calculate the value of their fitness measure. Evaluating a fitness measure at its time horizon usually requires prediction of both external environmental conditions (e.g., food availability, including the effects of competition, predation risk, weather) and the individual state variables used to represent fitness elements (e.g., size, energy reserves, reproductive reserves). These external and internal variables almost always depend on each other—e.g., future size and energy reserves depend on future food availability; future predation risk can depend on both predator density and individual characteristics such as size and activity. Therefore, we do not discuss prediction of environmental conditions separately from prediction of state variables.

We use the word *prediction* in part because that is exactly what we are modeling: the simulated individuals in our IBMs make explicit predictions of their fitness measure for each decision alternative they consider. But the use of this term also reminds us that our model individuals base trade-off decisions on predictions just like the ones we ourselves use every day: expectations that can be extremely simple and usually wrong, but still sufficient for us to make efficient and reasonably good decisions.

We discussed in section 1.4 the understanding in neuroscience that brains constantly make and use predictions to make decisions, which raises the hope that further understanding of brain function can lead to more accurate representation of prediction in IBMs. However, here our goal is not to design the most accurate models of prediction but to make models successful enough to produce realistic decisions. Simple models of prediction also have the advantage of computational tractability.

We address only methods that simplify the prediction sufficiently to allow fitness measures to be evaluated without employing complex calculation such as dynamic programming. All the methods allow expected fitness at the future time horizon to be calculated directly, although in some cases by using simulation of future events. What usually makes this possible is the simplifying assumption that if an individual selects an alternative, it will continue to use that alternative until the time horizon. This assumption is rarely accurate but very useful for producing good decisions because, in SPT, the individual updates its decision routinely: its actual future behavior can be quite different from the predictions it uses along the way to make decisions. But dynamic programming can also be used to evaluate fitness measures in SPT: Luttbeg and Schmitz (2000) and Luttbeg and Trussell (2013) developed simple SPT-like models in which individuals update behavior decisions as perceived predation risk changes by reexecuting a DSVM-like optimization.

In the following subsections we first provide general guidance and ideas on modeling prediction and then address prediction of specific fitness-related processes: growth, survival of starvation, survival of predation and other risks, and reproduction.

8.4.1 General Guidance on Modeling Prediction

Even though people naturally understand that prediction is fundamental to decision-making—we make very few decisions without considering what might happen in the future—few models in behavioral ecology have explicitly included or addressed prediction. Our experience points to three criteria for selecting or designing models of prediction for SPT. A method for representing prediction should (a) produce useful decisions; (b) be realistic, considering the individual organisms' sensing and cognitive abilities; and (c) be computationally tractable, to allow population-level simulation experiments.

Here we present some general approaches to representing prediction that we have either used in SPT models or considered promising. Any SPT model is likely to use more than one of these techniques, and these are no doubt only a small subset of ways to model prediction that deserve exploration.

While there appears to be sparse literature on explicit use of prediction in decision modeling, at least in ecology, there does seem to be growing interest in understanding how real organisms use prediction-like mechanisms. The last three general approaches addressed in this section concern use of memory, environmental cues, and updating based on experience; these techniques are especially exciting because they link new research on cognition by real organisms with population modeling. We cite examples of the literature on this topic but do not attempt to review it completely, in part because the review would quickly be out of date.

Assume that current conditions or rates persist until the time horizon. All our example models in previous chapters use as one of their prediction methods an extremely simple prediction of future conditions: that the conditions occurring when a decision is made will persist unchanged until the time horizon. This simple approach (used also by Luttbeg et al. 2003) has proved very useful even though the prediction is almost never correct. For example, the trout (section 2.2) and generic foragers (section 4.2) select habitat by assuming that the predation risk and energy intake rate they currently experience in each potential location would persist until their time horizon. While these assumptions seem very simple, their implications are actually complex and contradictory. For predation risk to stay constant over future time, not only would all the environmental variables affecting risk (for trout, these include water depth, velocity, and temperature) have to stay constant, but so would individual size if risk is size-dependent. But the other assumption, that the energy intake rate stays the same, implies that the individual's size *would* change due to growth. Despite these apparent problems, this simple prediction yields useful and realistic decisions. In fact, we recommend making this the default prediction approach and replacing it with something more complex only if it proves clearly inadequate.

(We have not explored one even simpler prediction: assume that future environmental conditions are always the same, regardless of current conditions. This approach may be useful in situations such as when conditions are highly variable in the short term but stable over typical time horizons, and when adapting to conditions has a high cost.)

Assume that recently experienced rates persist until the time horizon. Sometimes it is useful to represent prediction by assuming that recent experience, not just current conditions, will persist into the future. The life history decision in the *O. mykiss* facultative anadromy model (chapter 7) exemplifies this approach. For this decision of whether to commit irreversibly to becoming anadromous or remain resident in freshwater, model fish predict future probability of survival and future growth by "remembering" the survival probability and growth rate they experienced over the past 30 days and assuming that future survival and growth until the time horizon will equal the mean of the rates in memory.

This prediction method differs from the simpler approach discussed above in two ways. First, it averages conditions over a preceding period instead of assuming that conditions on the current time step will persist. This makes the decision less dependent on short-term variation in habitat conditions and interactions with other organisms, both of which can be significant in many environments. Considering recent trends instead of immediate conditions may often be appropriate for irreversible life history decisions. In contrast, using current conditions for short-term behaviors such as daily habitat selection often makes sense because individuals can change their decision as conditions change.

The second important feature of assuming recent experience will persist is that predictions and decisions depend on individual behavior, not just on what conditions were available. The survival probabilities and growth rates in the individual's "memory" are functions not just of past habitat conditions but also of what habitat the individual selected and its interactions with other individuals. Therefore, a short-term adaptive behavior (e.g., habitat selection) affects long-term behavior such as life history decisions.

Recent experience can be used for predicting future conditions in several ways. In addition to the simple approach in the *O. mykiss* model described above, another possibility is to project trends from recent experience by statistical modeling. Model individuals could be programmed to record the survival probabilities they experience and then use regression to predict future survival as a function of time or size. The Bayesian updating approach discussed below offers another way to base predictions on recent experience.

Use probability theory. Modelers often need to predict conditions that can be represented as probabilities. Survival of predation or other risks provides an example. If we assume an individual predicts that the probability of surviving the current time step t will persist until the time horizon T, the predicted probability of surviving until T is S^{T-t}. Of course the time units of S and t must be consistent: if t is in days, then S must be the probability of surviving for 1 day.

If the prediction is that S changes each time step (e.g., because the risk of predation changes as the individual grows), then we may need to predict the probability of surviving until the time horizon by determining its value on each time step until T and multiplying these probabilities together. Depending on how S changes, there may be useful approximations to reduce the number of computations, as discussed below for prediction of survival of starvation.

While most ecological modelers have experience modeling survival, other potential applications of probability theory to the prediction problem may be less familiar. Examples could include predicting whether an upcoming winter will be severe, the number of days without rain between now and the time horizon, or the number of predator attacks per day. Probability theory can often provide a powerful and flexible framework for modeling such predictions. Our suspicion is that many ecologists have had limited experience using the distributions most useful for modeling such problems, especially the discrete distributions.

The fishing fleet model of Yu et al. (2013) provides an example. The individuals in this model are fishing vessel operators who must decide before they leave port whether to use gear targeting swordfish or tuna. To make this decision, they predict the profit they would make using both gear alternatives. Regulations to protect endangered sea turtles affect the decision: the vessels are subject to a rule (based on actual regulation of the longline fishing fleet in Hawaii) that all fishing for swordfish must stop after a certain number of sea turtle "interactions," such as entanglements in swordfish gear. When we built the model, this quota was 22 sea turtle interactions per calendar year for the whole fishing fleet.

At the time fishing vessel operators make their decision, they know the value of the quota, the number of sea turtle interactions so far in the current year, the date, and the dates on which they expect to fish. The prediction they need is the probability that the interaction quota will be exceeded before the end of the trip they are planning. The Poisson distribution provides an excellent model for this prediction. It describes the distribution of discrete, independent, random events (here, sea turtle interactions) given a known mean rate (e.g., interactions per day). Straightforward computation from the distribution yields the probability that n events happen within a time interval t. The model vessel operators make the simple prediction that the average rate of turtle interactions so far in the current year (number of interactions divided by the current day of the year) will persist through their next trip. Then, from the cumulative distribution function of the Poisson distribution, they calculate the probability that the turtle interaction quota will be met before the end of their trip. (The prediction might be more accurate if turtle interactions per vessel-day of swordfish fishing were used as the Poisson distribution's rate parameter, instead of interactions per calendar day. But the model developers decided that per calendar day is more likely the way real vessel operators think about the interaction rate because they do not have easy access to information on vessel-days of swordfish harvest.)

This relatively simple prediction model results in a variety of realistic, complex, and nonlinear behaviors. When incorporated in the fishing gear decision, this choice between fishing for tuna or swordfish automatically depends on (a) the number of turtle interactions so far in the current year; (b) the date—e.g., if 10 interactions have occurred so far in the year, the interaction rate, and therefore the probability of the quota being exceeded during a trip, depends on whether the current date is early versus late in the year; and (c) trip duration—longer trips produce higher probability of the quota being met.

It is easy to envision other predictions needed in ecological IBMs that can be represented well using probability distributions. Familiarity with common discrete distributions is likely to benefit the formulation of predictions using SPT.

Use Bayesian theory to base predictions on experience. Bayesian probability theory includes updating of probability estimates as additional information is obtained, so its potential application to modeling prediction in IBMs seems obvious. *Bayesian updating* is a specific technique for using experience of discrete events to update prior probability estimates. The usefulness of Bayesian updating for modeling animal behavior has been widely explored (e.g., Valone 2006), and it may also be applicable to plants. An obvious potential application is updating the estimated daily survival probability used to predict future survival of predation, depending on whether the individual detects a predation event or observes a predator each day. If such an event were observed, estimated survival probability would go down and otherwise would go up. The simple updating algorithm is presented in Bayesian statistical texts such as Stauffer (2008) and spelled out for use in IBMs by Railsback and Grimm (2019).

We do not know of any examples of using Bayesian updating with SPT in an IBM, although Zhivotovsky et al. (1996) and Luttbeg and Trussell (2013) provide examples of combining Bayesian-like updating and optimization of expected future fitness to explore how predicted future conditions affect behavior. Updating probability estimates appears to have the potential to represent a form of prediction sophisticated enough to consider a type of learning. McNamara et al. (2006) review related approaches for applying Bayesian concepts to modeling behavior and the empirical evidence supporting them.

Model the mechanisms. The second version of our *Daphnia* model (section 5.4) assumes that model *Daphnia* predict the number of offspring that they would produce by the time horizon by executing an internal model that calculates their growth and size, and the resulting reproduction and survival probability, over time. This technique of assuming individuals use a simplified internal model to predict their future state was also used in the limpet model of chapter 6. The internal model used for prediction can be based on the same methods used to simulate what actually happens to the individual (e.g., what growth, size, and reproduction the simulated *Daphnia* actually experience after making their habitat selection decision), but it can also use a simplification of those methods. Simplification is beneficial to make the predictions computationally tractable— remember that individuals typically need to make many predictions to compare the available decision alternatives—but also perhaps to make the model more realistic than assuming individuals have perfect ability to foresee their future state. Sections 8.4.2 through 8.4.5 include examples of this approach applied to specific fitness elements.

Represent memory of time and space. Growing evidence indicates that some organisms use memory of past habitat conditions in their adaptive decision-making (e.g., Bailey et al. 1996), and behavior models that include spatial memory are now being explored. The third version of the *Daphnia* model (section 5.5) provides a simple example of using (nonspatial) memory, as individuals base decisions on survival probabilities and growth rates remembered from the previous day or night. We also used this technique in the version of the trout model that includes diel shifts in behavior (Railsback et al. 2005).

However, more sophisticated predictions that use spatial memory have been developed and tested. Amano et al. (2006) tested several alternative behavior submodels, including some with spatial memory. They found their goose foraging model better matched field observations when it did *not* assume geese use an accurate memory of resource availability over space but rely on social information instead. Bonnell et al. (2013) also tested alternative behavior submodels with different types of spatial memory, in a primate foraging model, and Kułakowska et al. (2014) did likewise in an IBM of foraging woodpigeons. These studies demonstrate the feasibility of representing spatial memory as a prediction method in IBMs.

Use environmental cues. Empirical research on how organisms use environmental cues to anticipate habitat conditions may contribute to realistic assumptions about prediction for use in SPT. Much of the field of "plant behavior" has focused on use by plants of chemical and light cues to anticipate competition by neighboring plants, herbivore attacks, etc. Animals also use environmental cues to alter their behavior in anticipation of future events. Antonsson and Gudjonsson (2002) found evidence that salmon in Arctic rivers use water temperature as a cue to the conditions they will experience upon migration to the ocean, with different responses to temperature in different rivers. Several studies (e.g., Boutin et al. 2006; Bergeron et al. 2011) indicate that squirrels and chipmunks use cues—perhaps chemical or visual signals from flowering—to "predict" high seed production by trees, and produce more offspring in anticipation.

Use intentionally inaccurate prediction to represent maladaptation. Sometimes the purpose of a model includes predicting the population effects of exposing individuals to conditions that they are not adapted to. Dill (2017) identifies as examples animals that (1) perceive ecotourists as predators, (2) do not recognize human hunters or fishers as a risk, and (3) use cues to predict resource availability (e.g., penguins who use cold temperatures and high chlorophyll as predictors of high prey abundance) when climate change has unlinked the cue from the resource. We can predict population effects of maladaptation by using models of prediction that reflect it. Our experience with this approach was a version of inSTREAM that represented how trout raised in a hatchery interact with and affect populations of wild trout. Hatchery trout were represented by modifying their predictions of future fitness at alternative locations to reflect hatchery conditions of uniform hydraulics, high fish densities, and abundant food: they were assumed unaware of competition, of how predation risk varies among habitats, and that exploration could result in finding better habitat. (In simulations, these virtual hatchery trout tended to displace the native fish but die before causing strong population effects.)

8.4.2 Predicting Growth and Size

Growth and size might not be considered fitness elements per se, because they do not necessarily contribute directly to fitness. However, in many models growth is important because it drives survival of starvation or predation and because growth and size strongly affect reproduction (or represent it, as in the trout model where growing to adult size represents reproduction in the fitness measure). Therefore, we often need to predict growth as part of SPT.

Most of our experience has been with one approach: assuming that the growth rate occurring under current conditions persists until the time horizon. This assumption treats any general patterns in growth that we may know about (e.g., that it increases or decreases with size) as unimportant between the current time and the time horizon.

Our approach tracks absolute growth per time step, e.g., grams of growth per day. Therefore, predicted weight at the time horizon W_T is calculated from current weight W_t and the absolute growth rate g_a that the individual would experience under current conditions if it chose the decision alternative being evaluated: $W_T = W_t + (T - t)g_a$. We have briefly explored use of a relative growth rate (grams of growth per gram of body weight per time), in hopes that its use would be an easy way to assume that absolute growth increases with size. In our experience the relative growth rate is less useful because its predictions of future size can be extremely sensitive to the value of the growth rate and often nonsensical. However, there may be little difference between the two approaches for short time horizons.

The *Daphnia* model of section 5.4 uses a subtly different approach: instead of assuming that the current growth rate remains unchanged until the time horizon, it assumes that current feeding conditions will remain unchanged and models the resulting growth and individual size over future time steps until the time horizon. Feeding conditions in the *Daphnia* model are represented simply by temperature, but other models may use more direct measures of growth conditions, such as the availability of food or feeding habitat. This approach seems likely to be most appropriate when external factors strongly limit food intake such that it changes little as individual state changes.

(Very simple models often assume that gross energy intake and growth are proportional, but this assumption is often not useful in IBMs with trade-off decisions. Often the trade-offs include factors that affect metabolic costs; examples include activity, temperature or other weather conditions, and whether to live in groups that cooperatively thermoregulate. Further, metabolic costs are generally assumed to vary nonlinearly over time with factors like individual weight. Whenever metabolic costs vary, gross energy intake can be a poor measure of growth or net energy intake.)

These approaches use what may seem like painful simplifications, but their predictions of future energy intake and size can suffice to produce useful adaptive trade-off decisions via SPT.

8.4.3 Predicting Starvation Risk

While ecology includes a long tradition of modeling growth and reproductive potential, much less effort has been expended on modeling mortality, particularly death by starvation. The lack of models of starvation may reflect the assumption that starvation is unimportant because ecologists rarely observe it. But when we think about behaviors that trade off energy intake against survival of predation, the importance of starvation becomes apparent. We rarely find starved organisms because, when faced with starvation, they take risks to avoid it and instead often end up being consumed themselves. Therefore, as with DSVM, many applications of SPT demand consideration of starvation.

As illustrated in section 4.5, starvation survival can have strong effects on ecology at the individual and population levels. Species and individuals that can resist starvation for long periods, like many fish (e.g., Simpkins et al. [2003] fasted juvenile trout for 147 days with little mortality), have great scope to reduce predation risk by feeding in safer but less productive habitat, feeding at safer times, or simply hiding. Therefore, populations of such species are buffered against short-term food shortages or spikes in predation risk. At the opposite extreme, animals such as small birds can survive at most a few days without feeding and thus have limited scope to reduce feeding to avoid risks other than starvation.

Because of the importance of potential starvation to many adaptive trade-off decisions, we need ways to model how individuals predict future starvation survival that are more sophisticated than the simple binary approach often used in DSVM—that starvation happens if and only if energy reserves fall below some threshold. This binary approach has two major disadvantages. First, as discussed in section 4.5, it does not give individuals the ability to avoid high predation risk by accepting a high risk of starvation when necessary; instead, it produces zero expected fitness for any alternative producing low energy reserves, while even extreme predation risk typically produces a low but still positive value of expected fitness. Second, the binary approach does not give individuals the ability to select the best among bad alternatives: when no alternatives provide sufficient energy intake to meet the starvation threshold, all the alternatives have the same zero value of expected fitness, and the individual has no ability to select the one offering the best chance of surviving until conditions improve.

Starvation, including susceptibility to disease and various syndromes resulting from inadequate energy intake, is a complex process (e.g., Simpkins et al. 2003), but it is typically modeled using a simple energy balance framework. (Here, we still use the terms "net energy intake" and "growth" interchangeably, assuming that all gross energy intake left over after metabolic costs is used for weight gain. This assumption is not valid for more complex representations of individual energetics that, for example, partition net energy gain into growth and reproductive reserves.) The forager model of chapter 4 and the trout model of chapter 2 are at the extremes of complexity in representing individual energy balances in IBMs, but even the trout model's representation of energetics is highly simplified. It represents how food intake, swimming speed, and temperature affect growth and starvation potential, because all these variables can differ among the habitat alternatives available to model fish. But the energetics model neglects other mechanisms that we know affect trout growth, such as how individuals allocate energy intake among somatic growth, lipid storage, and gonad production, because we did not expect these mechanisms to have effects on growth strong enough to affect habitat selection behavior and therefore population dynamics. Determining what mechanisms to include in the energetics model that determines growth and starvation survival is likely to be one of the most important model design decisions. This step especially benefits from the pattern-oriented model design strategy outlined in section 8.2.

Because the trout model's method for predicting future starvation survival has proved useful and flexible without excessive complexity, we present it here as an example even though parts of it clearly make sense only for animals like fish that typically have energy reserves much greater than their daily metabolic needs. (Railsback et al. [1999] provide a more complete description, including some computational considerations.)

First, the daily probability of surviving starvation and associated disease is modeled as a logistic function of a physiological variable called *condition*, represented here as K. We define condition as the individual's weight divided by the weight of a healthy individual of the same length. In the model it ranges up to 1.0, because if a fish of condition 1.0 grows in weight, it also grows in length. (Other assumptions are clearly possible; e.g., condition > 1.0 could represent additional energy reserves.) "Healthy" weight W_h is determined from length L using an empirical relationship $W_h = aL^b$, where a and b are found from regression on field observations. (To represent healthy individuals, this regression might exclude, for example, the 30% of observations with lowest weight per length.) This condition variable is one measure of an individual's energy status and has the advantage of being routinely measured in real fish. However, other measures of energy status could also be used to predict starvation survival.

The logistic function is

$$\text{Equation 8-1: } S_t = \frac{e^{(aK_t + b)}}{1 + e^{(aK_t + b)}}$$

where S_t is the probability of surviving starvation for the current time step, K_t is the current condition value, and a and b define the curve's shape. The starvation function is parameterized by assuming S_t is 0.1 when condition is 0.3 and 0.9 when condition is 0.6 (figure 8.2), which produces values of −6.59 and 14.6 for a and b. (Chapter 16 of Railsback and Grimm [2019] provides a simple recipe for parameterizing and programming logistic functions.)

It may seem generous to assume that an individual still has a 90% probability of surviving another day after losing 40% of its body weight, but keep in mind

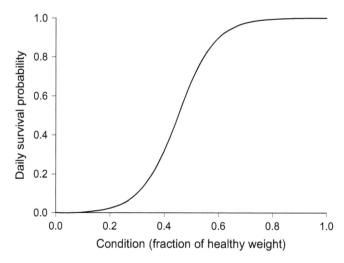

F<small>IGURE</small> 8.2. Logistic relation between starvation survival and condition in the trout model.

that fish lose weight gradually, even with no food intake; for condition to decrease to 0.6, the individual would have to survive many days of decreasing weight and decreasing survival probability. (We explore survival over long periods below.) An individual is actually unlikely to survive as condition declines from 1.0 to 0.6.

The second component of the starvation survival method predicts future condition for decision alternatives that differ in net energy intake. To make the approach represent how weight loss accumulates over time when net energy intake is negative, we do not use the simplest approach of assuming that the individual's value of condition under current conditions remains unchanged until the time horizon. Instead, we assume that the growth rate under current conditions remains unchanged until the time horizon, and then predict weight at the time horizon (W_T), as discussed in section 8.4.2 for predicting size. A simple algorithm determines condition at the time horizon: if W_T exceeds the weight of a healthy fish at the individual's current length, then the individual is assumed to have grown in length as well as weight, so its condition at the time horizon K_T must be 1.0. Otherwise (when growth is negative or insufficiently positive to overcome an existing weight deficit), the individual's length is assumed not to have changed at the time horizon, so K_T is calculated by dividing W_T by the weight of a healthy fish at the fish's length.

The final component is a way to calculate the probability of surviving starvation until the time horizon, considering the predicted change in condition and the logistic relation between condition and daily survival. The most accurate way of doing so is to iterate over each future time step, calculating the daily weight and condition and the resulting daily survival probability, then multiplying all the daily values together to get the probability of surviving until the time horizon. However, to reduce computation, we tried several approximation methods (Railsback et al. 1999) and determined that the first moment of the logistic function provides a reasonable approximation. We therefore approximate the probability of surviving starvation from the current time t until a time horizon T as

$$\text{Equation 8-2: } S_T = \left\{ \frac{1}{a} \ln\left(\frac{1 + e^{(aK_T + b)}}{1 + e^{(aK_t + b)}}\right) \Big/ (K_T - K_t) \right\}^{T-t}.$$

(When K_T equals K_t, which is very common when both equal 1.0, avoid division by zero by setting S_T to its value for K_t from equation 8-1, raised to the power $T - t$.) The resulting predictions of surviving starvation depend on the individual's net energy intake rate and current condition, as well as the time horizon (figure 8.3). The logistic function provides the useful characteristic that the probability of surviving starvation remains high for long periods of slight weight loss or for short periods of high weight loss, and then decreases rapidly as weight loss continues. Another beneficial characteristic of this approach is that while survival approaches zero asymptotically as weight loss accumulates, large *relative* differences among decision alternatives remain. For example, in figure 8.3 it appears that two individuals that start with $K_t = 0.8$ but experience weight loss of 0.5 and 1.0 g/d both have vanishingly small probabilities of surviving for 90 days. But in fact those probabilities differ by orders of magnitude (3.8×10^{-15} for 0.5 g/d weight loss vs. 4.6×10^{-100} for 1.0 g/d weight loss). These differences may seem

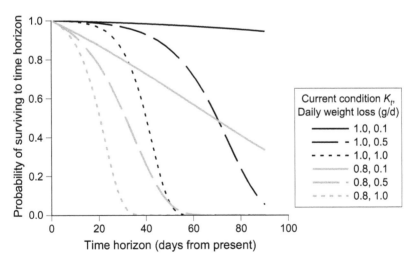

FIGURE 8.3. Probability of surviving starvation for a fish 20 cm long in the trout model, under different combinations of starting condition and rate of constant daily weight loss. The Y axis is the probability of surviving until the time horizon indicated by the X axis.

unimportant, but in fact they allow individuals to select the best alternative when none are good.

(It is critical when programming logistic functions to be aware of the potential to exceed the computer's precision, which can produce completely erroneous results in some programming platforms. This potential for "floating-point overflow" or "underflow" errors varies among programming platforms as well as with hardware. To explore this possibility, have your platform report the value of $\exp(n)^{T-t}$, where $T - t$ is the time horizon in equation 8-2, for values of n that increase until you get an error or erroneous result. If errors seem likely, they can be managed, e.g., by checking whether the numerator in equation 8-1 exceeds a threshold that causes the exp operator to produce a number larger or smaller than the computer's precision. For example, S_t can be set to 1.0 if the numerator is greater than 30 and to 0.0 if the numerator is less than -100. This check does limit the model's ability to distinguish among decision alternatives with extremely high or low expected survival.)

Finally, we offer an aside about the use of dynamic energy budget (DEB) theory (Kooijman 1993, 2010; Nisbet et al. 2000) in IBMs. DEB has been promoted as a general and theory-based approach to modeling individual energetics, so it is natural to consider it as a framework for predicting starvation survival for SPT. We recently had the opportunity to work with highly experienced DEB users to replace the energetics components of our inSTREAM trout model with DEB theory. The experience was not encouraging, because of well-known characteristics of DEB. First, its state variables are highly abstract and challenging to relate to measurable or clearly interpretable variables, even basic state variables such as length and weight. DEB also does not use an observable or easily interpretable measure of energy reserves. Second, DEB is simplistic in ways that sometimes prevent its use in IBMs: examples include how it represents feeding and food quality and the energy cost of activity. Because DEB was designed to be universal, it lacks the species- and life-stage-specific mechanisms we typically need to model adaptive behavior. More effort to develop and standardize energetics in models of growth and starvation survival in IBMs would certainly be useful. But in part because DEB was designed for purposes very different from SPT, using it to model growth and predict starvation in IBMs with SPT is unlikely to be straightforward.

8.4.4 Predicting Predation and Other Risks

Like starvation, predation risk has been studied and modeled less than growth and reproduction. We know from experience (e.g., Harvey 1991; Harvey and Nakamoto 2013; Harvey and White 2016, 2017) that quantifying predation risk and how prey perceive and avoid risk can be challenging both technically and administratively. However, indirect effects of predation risk on population dynamics and trophic interactions has been a major topic of ecological research for many years, and this work has clearly shown that we need models of risk, and perception of risk, to understand populations and communities. Our experience has been that we can often model predation risk sufficiently to produce realistic behavior via SPT using literature and mechanistic understanding along with limited observations.

We focus mainly on predation risk in this section, but the need to avoid other kinds of nonstarvation mortality can also affect behavior. In our trout model, habitat selection is driven by the need to avoid not just starvation and predation but also thermal stress and exhaustion from occupying habitat with excessive water velocity. The approaches described here can apply to many short-term risks.

Because it frees us of the need for mathematical optimization, using SPT allows more complexity and sophistication than DSVM in how we model predation and other risks. Our trout model, for example (section 2.2), represents two categories of predators, with daily survival probabilities for each kind calculated from multiple functions that represent how survival depends on both habitat variables (depth, velocity, proximity to hiding cover) and individual size. The model then predicts survival until the time horizon using the simple assumption that the survival probability occurring at the time of the decision, S_t, persists until the time horizon. Therefore, the predicted probability S_T of surviving predation until the time horizon is calculated as $S_T = S_t^{T-t}$.

Three of our example models illustrate another kind of realistic and important complexity that SPT can accommodate: variation in risk that happens cyclically at time scales much less than the time horizon. The version of inSTREAM with diurnal cycles (section 2.3 and Railsback et al. [2005]) and the *Daphnia* model of chapter 5 assume that predation risk is lower at night; the limpet model of chapter 6 assumes risk from both predation and desiccation occur only during unfavorable tides. Such cycles are likely to have important effects on adaptive behavior in many populations. To project expected survival of predation until the time horizon, these example models assume that the current survival probability at the *daily* scale persists until the time horizon. For a decision made at the start of night, a model trout calculates the daily survival probability (using a method described in section 2.3) as the combination of (a) survival probability presented by the night alternatives it is currently choosing from and (b) memory of the survival probability it actually experienced during the previous daytime.

When change in risk between the current time and the time horizon cannot be ignored, several alternatives are available. If a risk depends strongly on individual size or energy reserves, then the method described in section 8.4.3 for starvation should be adaptable. The second version of the *Daphnia* model (section 5.4) offers another approach: simulating growth and size using the same predictions and approximations used for other fitness elements and calculating the survival probability for each predicted future time step.

8.4.5 Predicting Reproductive Success

Of course population models that span the life cycles of the organisms of interest require some representation of reproduction. In addition, reproductive success is an essential component of fitness and therefore often has a central role in defining fitness measures in SPT. But what level of detail is needed to represent reproduction in a fitness measure? In addressing this question, remember that the goal of developing IBMs to estimate population- and perhaps community-level responses to environmental conditions differs from the goal of identifying optimal life histories assumed to be the products of evolution. While the literature, especially in life history theory, identifies many interesting complexities relating reproduction to fitness, our experience suggests that a relatively simple representation of reproduction can be adequate. Further, we recommend starting with a simple approach that might even leave explicit representation of reproductive success out of the fitness measure, and then using the theory development cycle of chapter 9 to determine how much detail is necessary to produce adaptive behavior useful for a particular population modeling problem.

Perhaps the simplest approach to predicting how reproductive success depends on behavior is exemplified by the forager patch selection model of chapter 4: the only requirement for potential future reproduction is to survive the current life stage.

Predicted future size can be a useful surrogate for reproductive success in fitness measures for organisms whose reproductive output is positively related to size, typically those exhibiting indeterminate growth. The original version of the inSTREAM trout model (section 2.2) represents reproductive success in the fitness measure only as progress toward reaching the minimum size for reproduction; once adult size is attained, size and reproductive potential no longer affect adaptive behavior. This simple approach produced useful and realistic habitat selection behavior in the original model version, but when we added activity selection as a second adaptive behavior (section 2.3) it was no longer adequate. With the additional behavior alternative of hiding instead of feeding, model trout unrealistically ceased growth as soon as they attained adult size: they were able to fine-tune their growth-risk trade-off to the point that they maintained only the minimal food intake necessary to avoid future starvation risk. Adding size as a surrogate for fecundity resolved this problem, providing an incentive for adult trout to continue growing (Railsback et al. 2005).

Future size is not likely to be a useful surrogate for reproductive success in organisms with determinate growth. Prediction of future reproductive output of mammals, for example, is likely to require consideration of growth to reproductive size and then availability of energy, above basal metabolic demands, for gestation, lactation, and parental care (e.g., Cook et al. 2001). At its simplest, consideration of these energetic demands could be represented as a required minimum level of energy reserves when adults commit to reproduction. More sophisticated approaches for predicting reproductive success often involve representing reproduction as an energy budget compartment. The *Daphnia* model of chapter 5 illustrates an extremely simple example of a reproductive energy compartment.

Reproduction often has important effects on survival and energetic status. When predicting consequences of decision alternatives on reproduction, it may also be important to predict costs of reproduction to survival of starvation and predation. These costs directly affect population dynamics and can affect adaptive trade-off decisions, such as whether and when to reproduce (Audzijonyte and Richards 2018). They can be represented simply in some cases: the inSTREAM trout model captures the survival and energetic costs of reproduction by having spawning fish incur a loss of body mass that increases the risk of starvation and leads fish to use habitat with greater risk of predation to recover their condition.

In one particular way, IBMs using SPT may be more complex than traditional life history theory that uses similar fitness measures: in an IBM, we usually cannot ignore males. For some models, it may make sense to give males and females separate fitness measures because the two sexes can exhibit different behaviors and costs of reproduction in many species and contexts.

8.5 STEP 4: SELECTING A DECISION ALGORITHM

The fourth step in SPT is designing the algorithm individuals use to decide which alternative to implement. This step is not part of DSVM, which assumes individuals follow the decision pathway identified via optimization. However, decision algorithms are the subject of an extensive literature in behavioral ecology and, especially, human decision theory.

The decision algorithm used in all the example models of this book is so straightforward that readers may not even recognize it as an explicit decision method: individuals evaluate all the feasible alternatives and select the one with the highest value of the fitness measure. This "select the best" decision algorithm is likely to be the obvious choice for many modeled decisions, but others may be more appropriate in some situations. The "select the best" approach assumes that individuals have the resources and ability to evaluate all the alternatives—we assume, for example, that our simulated trout can evaluate up to several hundred habitat cells within a radius of their current cell, because we know that stream fish invest some fraction of their time and energy exploring their surroundings (Harvey and Nakamoto 1999; Harvey et al. 1999). When it is reasonable to assume the individuals are familiar with all alternatives, then "select the best" is likely the simplest and most reasonable approach.

However, decisions that require substantial resources may create a trade-off between the costs and benefits of evaluating alternatives. Decision theorists understand and have modeled this trade-off, producing decision algorithms that either incorporate the cost-benefit trade-off explicitly or simply limit the effort used to make a decision. "Efficient" decision methods are often designed to reduce the number of alternatives that must be examined before selecting one. Satisficing, for example, refers to selecting a good enough alternative: an individual could examine alternatives sequentially and select the first one that provides fitness above some threshold level, or could examine a certain number of alternatives to establish an expectation of a good alternative and select the next one that meets the expectation (Gigerenzer et al. 1999). We have not yet found these efficient algorithms useful for our models, in part because they often turn out not to be as simple as they first appear: What should an individual do if no alternatives meet the satisficing threshold, or if none of the subset of alternatives that have been examined offers a good chance of future survival? The extensive literature on simple decision heuristics (e.g., Gigerenzer et al. 1999, Gigerenzer and Selten 2002) is also of limited value because many of the heuristics address how to make decisions without the detail needed to evaluate trade-offs—they may be appropriate for modeling how people make financial or social decisions without much thought but not for how organisms make highly evolved life-or-death decisions.

Computational limitations can be another consideration in selecting a decision algorithm, especially when modeling large populations. Given modern computing power, we generally recommend that the choice of decision algorithm be based primarily on biological considerations—modeling organisms and decisions using realistic assumptions—unless it becomes clear that computation is a real concern *and* that a more tractable approach produces usefully realistic behavior.

8.6 STEP 5: IMPLEMENTING AND TESTING THE THEORY

The final step in SPT is to implement it in an IBM that represents the individuals and all the mechanisms that drive their decisions and determine population-level consequences. For this step, we again refer readers to the existing literature, especially Grimm and Railsback (2005) and Railsback and Grimm (2019), for guidance on implementing, analyzing, and doing science with IBMs.

Implementation and testing of SPT should take place in two stages. First, the theory needs to be fully programmed and explored by itself—*before* putting it in an IBM. We normally program the entire decision submodel, including the prediction methods and even the mechanisms of how growth and survival depend on variables of the individuals and their environment, in a software platform such as Excel or MatLab. Implementing the SPT theory by itself is essential. Doing so reveals all the complications and ambiguities we did not think about when writing the first-draft verbal description of the theory, and allows us to fix any problems. This is also the stage at which submodels driving the theory can be calibrated and tested against quantitative data as needed. The independent implementation also allows us to thoroughly explore and evaluate the decision submodel by itself; without doing so, it would be (or *should* be) impossible to convince anyone of the submodel's usefulness, and impossible to understand how the complete IBM works. The decision submodel should be explored by examining the fitness measure's value under the full range of conditions that could occur in the IBM (figure 8.4).

The second implementation stage includes programming the full population model, providing the graphical interfaces essential for understanding and testing IBMs, and testing the software thoroughly. The availability of the NetLogo software platform (Wilensky 1999; Railsback and Grimm 2019) has greatly facilitated this step. We recommend NetLogo even for complex IBMs because it allows both new and experienced modelers to rapidly implement models and start learning from them. NetLogo has also proved efficient computationally, often providing better speed and support for large simulation experiments than either custom software or lower-level modeling platforms (Railsback et al. 2017).

Once an IBM is programmed, it can be used for the final stages of modeling that traditionally include calibration and validation against quantitative data, sensitivity and uncertainty analysis, and simulation experiments to address its original purpose. However, at the end of this chapter and in chapter 9 we discuss an additional task that, for IBMs of adaptive individuals, is often critical and the most likely to produce important discoveries.

8.7 CONCLUSIONS

What we call SPT is really a general process for modeling adaptive trade-off decisions given uncertain and unstable future conditions. SPT is especially intended for use in IBMs that represent feedbacks from behavior: when individuals must

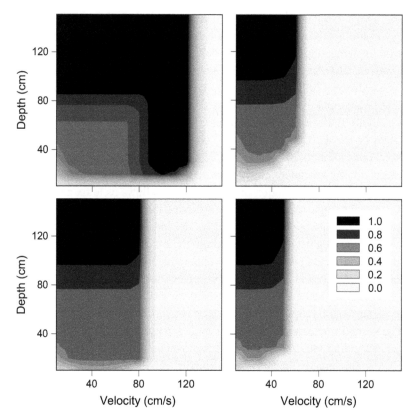

FIGURE 8.4. Example exploration of an SPT decision submodel: the fitness measure from the trout model of chapter 2. The contours indicate how the fitness measure—expected survival of predation and starvation to a sliding time horizon 90 days away—varies with stream habitat velocity and depth. Top and bottom rows, respectively, display values for typical age two and age one trout. Left and right columns, respectively, represent habitat with vs. without velocity shelter that reduces swimming speed and respiration costs. (From the exploration used by Railsback et al. [2003] to compare SPT to habitat selection modeling.)

make decisions affected by the behavior of others. In this chapter we have outlined a process for developing SPT for specific situations: selecting an appropriate fitness measure, modeling how individuals predict internal and external variables used to evaluate the fitness measure, and selecting a decision algorithm that the individual uses to choose among alternatives after evaluating the fitness measure for each of them.

SPT development normally nests within the overall process of developing the IBM that links individual behavior and population (or community) dynamics. Designing the IBM's structure—identifying the entities in the model, their state variables, the mechanisms that change the state variables, and the scales at which the model operates—is how we also complete the first critical stage of SPT: exactly defining the decision it models. The pattern-oriented model design strategy can efficiently determine what must be in the IBM and, of equal importance, what can be left out.

The initial steps in implementing SPT are similar to those of DSVM, but subsequent steps differ. Design of DSVM, SPT, and in fact all models should begin with careful specification of exactly what problem the model addresses. The second step, designing a fitness measure, closely corresponds in DSVM and SPT, and the DSVM literature should be very helpful to those using SPT.

The third step of deciding how to model prediction is unique to SPT. DSVM does not represent prediction of environmental conditions because it simply assumes that the individual knows what future environmental conditions will be and that those conditions are unaffected by behavior. DSVM does model how an individual's state changes over time but rarely considers the concept that SPT depends on: that individuals can use predictions simpler than reality to make good decisions. Knowing the future is less important for SPT because, unlike DSVM, it allows individuals to routinely reconsider and change their decisions as conditions change.

DSVM also does not include an explicit step of selecting a decision algorithm; instead, its optimization finds the best path through a fixed set of alternatives and assumes that the individual follows that optimal path. Most applications of SPT so far also assume that individuals evaluate a set of alternatives and select the one providing the highest value of the fitness measure, but it also allows other decision algorithms that may sometimes be more realistic or efficient.

The process we describe in this chapter may seem messy, especially to those most experienced with mathematical formalism. SPT differs from traditional ecological theory because our first priority is not mathematical elegance but finding theory that works in realistic contexts. The process of SPT development benefits from familiarity with the DSVM literature. However, it also depends on other skills and knowledge. The importance of familiarity with the organisms being modeled, including both their physiology and ecology, should be clear. (One eye-opening experience for us was starting to model a fish species that scrapes biofilm off rocks instead of drift-feeding for macroinvertebrates so, as in the limpet model of chapter 6, energy intake is controlled by digestion rate instead of food capture rate. This physiological difference completely changed the design and results of the SPT model of behavior.) We also point out the value of familiarity with at least the basics of probability theory.

The process of developing SPT benefits from thorough exploration and testing of the decision model's components as they are developed. Doing so inevitably reveals unforeseen complexities, problems, mistakes, and (often) surprising and realistic dynamics. It is absolutely essential to program the decision model independently and thoroughly understand its behavior before putting it into an IBM. Critical features of the decision theory can be difficult or impossible to discern when embedded with all the other dynamics in an IBM.

But once an IBM includes an effective decision model, links between individual behavior and population dynamics can be explored. This stage takes us from model development to science: using hypothesis testing and inductive reasoning, and confronting our theory with empirical observations, to illustrate and improve how well our implementation of SPT represents real individuals and systems. This is the final step of our SPT process and the subject of chapter 9.

CHAPTER 9

Testing and Refining State- and Prediction-Based Theory

9.1 INTRODUCTION AND OBJECTIVES

The main idea of SPT is that simple, approximate models of prediction and decision-making can produce behavior realistic enough to make individual-based population models useful. But this claim raises an obvious question: How do we determine whether behavior theory is "realistic enough" or an IBM "useful"?

Testing and refining the behavior theory in IBMs is especially important. In classical theoretical ecology, "testing" of theory might concern only how well a model could be fit to a long time series of population data, or might simply be neglected. But with SPT and IBMs, our goals include producing models that are sufficiently mechanistic to be useful when we do not have population time series data for calibration, for example, when we need to predict how a population responds to conditions we have never observed. Instead of relying on calibration of simple models to data, we rely on the models' mechanisms: we try to make our models represent the major processes that determine how the real population responds to the drivers of interest. To evaluate IBMs and develop confidence in their results, we need to test the reliability of the mechanisms in them, including, especially, the behavior submodels.

But establishing a model's credibility is not the only reason to test theory for behavior. Doing so also offers a new and productive approach to theoretical ecology: a way to develop a toolbox of across-level theory useful for modeling populations of adaptive individuals. We refer to testing and refining behavior submodels as *theory development*, and we do it by following the classic inductive reasoning cycle of posing, testing, and falsifying alternative hypotheses.

In this chapter we provide a brief introduction to the pattern-oriented theory development process and several examples. Grimm et al. (2005), Grimm and Railsback (2005, 2012), and Railsback and Grimm (2019) provide further discussion and examples of the approach. Inspired by the "synthetic microanalysis" of Auyang (1998), this approach is not limited to SPT but applies in general to the behavior theory we use in IBMs.

9.2 THE PATTERN-ORIENTED THEORY DEVELOPMENT CYCLE

The basic idea of pattern-oriented theory development for IBMs is simply to hypothesize several alternative theories—submodels—for the behavior of interest, test them by implementing them in an IBM and seeing how well the IBM reproduces characteristic patterns observed in real populations, and then "falsify" theories that do not cause the IBM to reproduce the patterns (figure 9.1). This cycle can be repeated, in conjunction with further empirical research, by revising the theories and testing them against new patterns that provide additional resolution. The cycle can be mapped in four steps.

Step 1: Identify characteristic observed patterns that will serve as the criteria for testing theory. These often include the patterns used to design the IBM, as we discussed in section 8.2. The patterns can be extracted from the literature or from new field observations.

Selected patterns should characterize the system with respect to the problem the IBM is designed to address. This means that the patterns should be driven by the same mechanisms believed to be important for the model's purpose. Our trout model, for example, was designed to predict the effects of flow and temperature

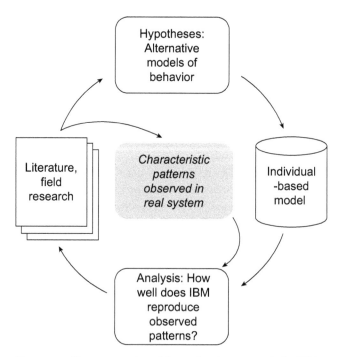

FIGURE 9.1. The pattern-oriented theory development cycle for IBMs.

regimes on populations. We assumed, from the extensive salmonid literature and our experience, that flow and temperature affect trout populations via mechanisms such as how habitat depth and velocity affect growth and predation risk, and that trout adapt to changes in flow and temperature via habitat selection behavior. Therefore, to test our theory for habitat selection, we looked for patterns in how habitat selection responds to changes both in temperature and in the processes that interact with depth and velocity to drive growth and predation risk.

Selected patterns should also *emerge* from the mechanisms in the model instead of being imposed, or hardwired, by model assumptions. For example, bigger trout using deeper habitat than smaller trout is a widely observed pattern driven by mechanisms of interest in our trout model. But our parameters for predation risk make deep habitat safer for large trout and shallow habitat safer for small trout, so the pattern is strongly imposed. But a pattern in how habitat selection changes in response to factors that affect growth (e.g., presence of competitors for food) or risk (e.g., presence vs. absence of fish predators) not only depends on parameter values but also emerges from how the trade-off between growth and risk is modeled.

Qualitative patterns are generally more useful than quantitative patterns. We try to conduct this theory development cycle before detailed calibration of the IBM, which makes quantitative model results uncertain. Further, qualitative patterns should be more robust than quantitative patterns to uncertainties such as measurement error. Most importantly, the whole cycle seems more convincing when based on clear and noncontroversial qualitative trends.

The variety of patterns tested also deserves consideration. Patterns should include responses to different model inputs or drivers. They ideally include patterns at both individual (e.g., how individuals change their behavior in response to some change) and population levels (e.g., how distribution, competition, abundance, or survival rate responds to some change). These patterns should also occur at a diversity of spatial and temporal scales, including, for example, patterns in how individuals interact with each other or their local environment over short times as well as long-term patterns in population responses to widespread physical or biological conditions. The examples provided in section 9.3 illustrate the use of patterns at multiple levels and scales. Using a variety of patterns tests the behavior theory's generality: its ability to predict responses to all the factors that could be important to the population dynamics of interest.

Step 2: Hypothesize alternative models of behavior. Alternative models are competing theories to explain the behavior of interest. Comparison of multiple theories makes the theory development cycle more informative and convincing. We need to hypothesize and test alternative theories, even those that we know will not work well, for several reasons.

It is valuable to start with an extremely simple "null" theory that assumes little or no fitness-seeking ability. Often, the null theory assumes random decision-making.

The null theory provides a baseline contrast for other theories and tests the possibility that the adaptive behavior being studied is actually not very important.

We also have found it informative to include classical theory from behavioral ecology to show how well it works in an IBM designed to address population problems. However, sometimes classical theory simply cannot be included in the theory development cycle. When we developed the original SPT for habitat selection in the trout model (Railsback et al. 1999), we tried to consider the theory that animals maximize their fitness by minimizing the ratio of predation risk to growth (section 1.3). In simulations of natural habitats with realistic temporal variation in environmental conditions, this theory produced impossible results. This outcome actually made sense because the assumptions under which the theory was developed (Gilliam and Fraser 1987) are not met in our IBM. All theory in behavioral ecology uses simplifying assumptions, and the underlying assumptions in classical theory often do not apply in IBMs that include natural complexities. It can therefore be valuable to use the theory testing cycle to explore the conditions under which classical theory does or does not work.

The alternative behavior theories should of course include the SPT that was designed specifically for the IBM, but often can include several versions. Our *Daphnia* and limpet models each developed and explored several versions of SPT that included increasingly accurate prediction methods (sections 5.3–5.5, 6.4–6.5). Those explorations were, in essence, pattern-oriented analyses that tested for benefits of additional model complexity.

Step 3: Implement each hypothesized theory in the IBM. The IBM's software should be thoroughly tested before the analysis of alternatives. However, we do not recommend careful calibration of the model at this stage. Instead, we use our best precalibration estimates of parameter values. Calibration of course depends on the specific behavior submodel being used, and recalibration of the IBM for each alternative introduces the potential for bias. If the analysis of alternative behavior theories is based on qualitative, not quantitative, patterns, then calibration should not be important to it.

Step 4: Analyze how well the IBM reproduces the observed patterns. Simulation experiments are designed and executed to test whether the model reproduces each pattern, under each hypothesized behavior theory. Theories that fail to reproduce patterns are then "falsified" as inadequate models of the behavior.

While the analysis needs specific criteria for whether each pattern is reproduced, it may benefit from limited emphasis on quantitative results. Ecologists are trained to contrast experimental treatments using statistical tests, but for the contrast of alternative behavior theories we find it more convincing to use simple, qualitative, and often graphical comparisons of results, combined with exploration and explanation of what happened in the model to cause different outcomes under different theories. One reason for preferring a qualitative approach is that many scientists are appropriately skeptical of statistical analyses of simulation results, because statistical conclusions can easily be influenced by simulation arti-facts and the number of "replicates." More importantly, an understanding of why the model produced the general patterns it did is much more informative than statistical outcomes. As Platt (1964) suggested: "Equations and measurements are useful when and only when they are related to proof; but proof or disproof comes first and is in fact strongest when it is absolutely convincing without any quantita-tive measurement."

However, when quantitative approaches make sense, statistical comparison of model results to empirical observations can be readily accomplished. Standard goodness-of-fit statistics may suffice in some cases; randomization and bootstrap tests (e.g., Manly 1997) often are more appropriate. In addition, Piou et al. (2009) have developed methods analogous to model selection (e.g., Akaike's information criterion) applicable to pattern-oriented evaluation of IBMs. Regardless of the methods used to make statistical contrasts, evaluators should focus on the eco-logical significance of the differences involved. As modelers seek to identify the optimal level of model complexity, small differences among alternative models in their fit to one particular set of observations should not be overemphasized, even when they can be statistically distinguished.

Step 1 again: Iteration of the cycle. Finally, the entire theory development cycle can be repeated if the first iteration fails to distinguish sufficiently among competing hypotheses. (The cycle might also be repeated to find theory for a different but related behavior or for a new set of conditions that affect behav-ior.) The cycle restarts with a return to the field, laboratory, or literature to iden-tify additional patterns. As some of the following examples show, the cycle can become an explicit way to link empirical research with theory development: new field experiments can be designed specifically to identify patterns that distinguish among competing hypotheses for behavior, which then improves the IBM, which can then make predictions that can be tested with field experiments. Chapter 11 further explores the theory development cycle as a framework linking modeling and empirical research.

9.3 EXAMPLES OF THEORY DEVELOPMENT AND TESTING

9.3.1 Literature Examples

A number of publications in ecology and other fields have tested how well an IBM reproduced a variety of observations under different assumptions about indi-vidual behavior. Many of these examples were produced before our first attempt to encourage pattern-oriented analysis of IBMs as a general approach to theory development (Railsback 2001) and our first published example (Railsback and Harvey 2002). Prominent examples of pattern-oriented theory development in individual-based ecology include

- Fish schooling and bird flocking behavior models, which have been tested against observed patterns in an extensive literature, from Aoki (1982) to Huth and Wissel (1992) to Ballerini et al. (2008) to Hemelrijk and Hildenbrandt (2015).
- Models of how trees adapt their shape in response to local availability of resources (e.g., Umeki 1997).
- Theory for interference competition in cranes and shorebirds and how it affects populations (Stillman et al. 2000, 2002; Stillman and Goss-Custard 2010).
- A contrast of alternative assumptions about how goose flocks select feeding habitat (Amano et al. 2006).
- A contrast of alternatives for three dimensions of foraging movement behavior (memory type, memory size, and social rules) in primates (Bonnell et al. 2013).

Ainara Cortés-Avizanda's elegant study of griffon vulture communication (Cortés-Avizanda et al. 2014) is a particularly noteworthy (though non-SPT) example of this way of doing ecology. Understanding how these birds find and aggregate at carcasses is no doubt essential for managing their populations and the populations of other birds and wildlife that also depend on carcasses (e.g., Cortés-Avizanda et al. 2010, 2016). Her simple IBM contrasted three alternative hypotheses for how vultures find food, with and without using information from other individuals. The patterns used to contrast the hypotheses were observed in a field experiment designed specifically for that purpose. The experiment showed that communication is essential to modeling how vultures find food, but also disproved a long-held assumption about *how* they communicate.

9.3.2 Trout Habitat Selection

Chapter 2 introduced the first explicit theory development exercise we conducted: the analysis of habitat selection theory in our trout model (Railsback and Harvey 2002). After developing an SPT representation for how trout trade off growth and predation risk in habitat selection, we contrasted that theory with two simpler alternatives. The first alternative has probably been the most commonly used decision objective in behavioral ecology: maximizing current growth rate. The second is just as simple: maximizing current survival probability, included because our IBM explicitly includes predation risk and its effect on behavior. We identified six patterns of how habitat selection in trout has been observed to respond to a variety of factors and events that affect growth and condition, mortality risk, or both.

As expected, both the SPT theory and maximizing growth caused the IBM to reproduce three patterns driven by factors that affect growth. The first pattern is that during flood flows, adult trout simply move from the main channel to lower-velocity margin habitat until the flow subsides; the extreme velocities during a flood make food difficult to catch and require high metabolic costs to maintain position, and hence cause negative growth. The second pattern is that trout feed in a size-based hierarchy, with larger individuals excluding smaller ones from otherwise better habitat. This pattern was reproduced because the model assumes fish have access only to food not consumed by larger fish in their cell, so the presence of larger fish can result in low growth. The third pattern is similar: that the introduction of a larger competitor (e.g., an introduced species) causes a shift in feeding location, again due to competition.

Also as expected, both the SPT theory and maximizing current survival caused the IBM to reproduce two patterns that concern survival. Maximizing current survival reproduced the first pattern (avoiding flood flows by moving to the stream margin) because swimming against flood flows poses a risk of mortality via exhaustion. Both SPT theory and maximizing survival also reproduced a fourth pattern: that the introduction of predatory fish causes juvenile trout to select shallower and faster habitat. In this scenario, the risk of predation shifts from being dominated by terrestrial animals, avoided by using deeper water, to being dominated by predatory fish, avoided by using shallow habitat.

Neither maximizing current growth nor current survival reproduced the last two patterns. Patterns five and six are that trout shift to higher-growth habitat at higher temperatures and that they alter their habitat selection quickly when food availability severely declines—long before starvation is imminent. Only SPT reproduced these two patterns because they both involve future, not immediate, consequences of a decision. Higher temperatures increase metabolic rates and the amount of food needed to avoid starvation, so both of these patterns are responses to a reduction in growth. Reduced growth and weight loss pose negligible risk of *immediate* starvation (fish can typically survive for months without food), but when decisions are based on survival until a time horizon, in this case 90 days in the future, starvation becomes an important part of the decision and fish respond rapidly. The main lesson from this theory development process was therefore that modeling the trout habitat selection decision in a way that produces behavior realistic enough for the IBM's design purpose requires theory that considers growth and predation risk *and* the future consequences of each day's decision.

9.3.3 Activity Selection in Trout

Chapter 2 also introduced our second theory development cycle. We used it upon adding a second adaptive trade-off behavior—activity selection—to the trout model, so that the individuals decide whether (and where) to feed or hide during the day and night (section 2.3 and Railsback et al. [2005]). In this case, we did not contrast our SPT submodel with alternative theories because we could not identify any alternatives that seemed likely to be useful or informative.

The activity selection analysis benefited from a recent burst of empirical research on activity selection as an adaptive trade-off behavior in salmonids (e.g., Metcalfe et al. 1998, 1999). We were able to show that our SPT theory caused the model to reproduce six observed qualitative patterns in diurnal vs. nocturnal feeding and how the prevalence of either mode depends on environmental conditions (temperature, flow, food availability, and competition) and individual state (fish age, condition, and life stage), and in how habitat selection differs between day and night. This analysis showed that our SPT theory for the two contingent adaptive trade-off behaviors, while complex and challenging to implement, could produce realistic emergent dynamics at both the individual and population levels.

9.3.4 Foraging Habitat Selection in Songbirds

The third example that we participated in developed foraging theory for an IBM of pest control services provided by songbirds in coffee farms (Railsback and Johnson 2011, 2014). This analysis did not involve SPT because the model did not include predation risk as a spatially variable factor driving habitat selection. Instead, birds were assumed to reduce risk by foraging as efficiently as possible to minimize the time spent feeding. The theory addressed how birds select foraging habitat within a grid of 5 × 5 m patches, at time steps as short as 1 minute, when patches vary in the food intake rate they provide to birds. Further, the food in each patch was assumed to be depleted over each day, so the birds compete indirectly for food. A variety of patterns observed in real coffee farms by our colleague Matthew Johnson and his students informed the analysis.

In selecting the alternative foraging theories to test, we followed the advice in section 9.2 and included a range of theories, from extremely simple (assuming birds move randomly) to unrealistically smart (assuming birds can detect the patch providing highest food intake, over a radius of 100 m). We also followed the above advice and attempted to include a classical theory of behavioral ecology, the *marginal value theorem* (Charnov 1976). Under that theorem's assumptions, an individual maximizes its food intake by staying in one patch until that patch's food availability is depleted down to the landscape-average level.

The coffee farm analysis provided another illustration of how classical theory of behavioral ecology can be inapplicable in an IBM. The marginal value theorem could not be used in the model for two reasons. First, it does not answer the right question: we need theory for which patch to move to, but the theorem only determines when a bird should leave its current patch. Second, "landscape-average food availability" is not a useful concept in a landscape full of other birds depleting the food. Further, much of the coffee farm landscape provides poor foraging, so the landscape-average food availability is insufficient to maintain birds (or we are forced to somehow delineate the habitat that birds do and do not consider as part of their foraging landscape).

The performance of the null hypothesis was noteworthy in this example. Random habitat selection caused the IBM to at least partially reproduce six of the nine patterns used to test theory. We can infer from this outcome that too simple, or too few, patterns may not have the power to identify useful theory.

One final key result from this example was rejection of the hypothesis that assumed birds can identify the best patch over very long distances, because it did not reproduce the shape of an observed distribution of movement distances. This result illustrates that we can easily develop theory that overestimates the capabilities of organisms, and that the pattern-oriented theory testing process can identify and exclude theory that inadequately reflects limitations of the sensing and cognitive abilities of individuals.

9.4 CONCLUSIONS

The simple and straightforward pattern-oriented theory development cycle (figure 9.1) will be familiar or unsurprising to most scientists. Even so, we think that the cycle could be the most important idea in this book, a way for ecology to move more rapidly toward being a science with theory useful for solving the real-world management problems that practicing ecologists face every day. Testing theory in this way yields the across-level theory discussed in chapter 1: models of individual behavior that are proved useful for predicting population-level phenomena. It is also how we move past one of our main concerns with the historic practice of individual-based modeling: the widespread use of IBMs that contain individual adaptive behavior without showing that the simulated behavior is sufficiently realistic to be useful.

Hypothesizing and contrasting alternative theories is a key element of the theory development cycle. Many published "tests" of theory in ecology, including behavioral ecology, look at how well predictions of one theory correspond with empirical data, but few studies test alternative theories and try to distinguish which one best corresponds with observations, especially observations selected specifically for their power to falsify alternative theory. Ecological theories cannot be proved "true," but we can reject theory that clearly is not useful in specific contexts for specific problems. And we may find that useful theory is simpler than we anticipated: the bird foraging theory that worked best to model pest control services in coffee farms simply assumed that birds can, within one- to several-minute time steps, select the patch offering the highest food intake rate from among the nine patches in their immediate neighborhood (Railsback and Johnson 2011).

This separate modeling step of testing theory for behavior is unique to individual-based modeling. It is tempting to think of it as a form of model validation, but we consider it instead an essential part of designing IBMs, especially IBMs of individuals making adaptive trade-off decisions. But once theory development is completed, modelers can move on through the more traditional late stages of model development, such as calibration and sensitivity and uncertainty analysis, with guidance for IBMs by Grimm and Railsback (2005) and Railsback and Grimm (2019). Validation against observations is often considered the ultimate stage of model development, and, through our experience modeling trout, we have learned that validation too is different when models include trade-off behavior. In the next chapter we discuss how models of populations of adaptive individuals gain credibility.

CHAPTER 10

Building Model Credibility

10.1 INTRODUCTION AND OBJECTIVES

In modeling populations of adaptive individuals, we usually model questions about real populations or communities. This is not just because IBMs with adaptive behavior are often the only practical tool for predicting how ecological systems respond to management alternatives. Even when doing "theoretical" ecology, working on a clearly specified problem of a real system makes it easier to design and test a model because it provides observed patterns for designing and testing theory for adaptive behavior as well as a clear basis for model assumptions. Working on real problems in real systems can also heighten the need to address model credibility. Of course potential users of models want to know how much confidence to place in them. SPT, as a nontraditional approach to modeling adaptive behavior embedded in a nontraditional population modeling approach (mechanistic individual-based modeling), faces a significant credibility challenge. This challenge is complicated by the many ways that models can gain or lose credibility, and widespread confusion surrounding the term *model validation* (e.g., Augusiak et al. 2014). The ability to be tested against many kinds of observations has long been claimed an advantage of IBMs, but with that advantage comes the increased risk of a model being inaccurately "invalidated" by observations that are not truly comparable to model results.

In this chapter we address the task of testing, improving, and establishing the credibility of IBMs that contain adaptive individual behavior. Our experience with trout and salmon models provides the primary basis for this discussion, but other long-term modeling projects have produced similar experiences (e.g., Shugart et al. 2018; Stillman and Goss-Custard 2010). We begin by summarizing some of the issues and challenges that typically arise and how they have been dealt with, and then present lessons learned from two decades of empirical and simulation studies addressing credibility of the salmonid models.

10.2 ISSUES IN "VALIDATION" OF
INDIVIDUAL-BASED POPULATION MODELS

On many occasions, we have faced the demand for "model validation" from prospective users and others with interest in model output. These demands usually seek evidence of a match between model output and independent datasets that cover fundamental responses, such as annual variation in population abundance. Many people consider this form of evidence, termed *predictive validation* by Rykiel (1996), necessary for consideration of any unfamiliar model.

However, a variety of problems have been attributed to this form of validation. Holling (1978) stated that such validation can add credibility but falls short of establishing the "truth" of a model. Holling also noted that matches between empirical observations and model output could occur by chance or for the wrong reasons. Rykiel (1996) pointed out that validation requires the specification of criteria for success, a rare step open to disagreement among evaluators.

In our experience, one practical problem for this form of validation is the rarity of useful empirical data on appropriate temporal and spatial scales, particularly datasets that include strongly altered environmental conditions that might provide the basis for a robust test of a model. The potential to predict outcomes under novel conditions can be considered the most valuable feature of IBMs utilizing SPT and other mechanistic models; but robust data from such conditions are unavailable by definition. In some situations, predictive validation may be precluded because predictions or information on temporal variation in key drivers of a model are not available. For example, as one might expect, baseline rates of predation and food availability strongly influence results of the trout model. Our observations of natural populations of fish on the scale of individual rivers or creeks suggest that currently unpredictable annual variation in predation risk (as influenced by events such as visits by groups of otters) can strongly influence population dynamics. However, our trout IBM uses a constant value for baseline predation pressure in the absence of any data on its annual variation or any ability to predict it. We also suspect that in some aquatic ecosystems annual variation in food availability may be significant, but we are in a similarly weak position to predict such variation and therefore treat the baseline value of food availability as constant across years.

Richard Stillman and his colleagues (Stillman et al. 2000; Stillman and Goss-Custard 2010) offer a solution to the problem for validation of unmeasured annual variation in key drivers of IBM results. They have compared the fundamental prediction of their IBMs for coastal birds—overwinter mortality—to empirical data by comparing the averages of predicted and observed annual results grouped by conditions important to the model results. For example, in a test of their oyster-catcher model, they grouped years by bird abundance because of known density-dependent overwinter mortality. Averaging predictions and observations over a set of years reduces the influence of unmeasured annual variation. In this example, Stillman and Goss-Custard offered the energy content of oystercatcher prey as an influential parameter that might have varied annually. Clearly, this approach relies on a substantial amount of data, but its end result is that the family of coastal bird IBMs has garnered a high level of credibility and acceptance through comparisons of their fundamental predictions to observations.

A second common problem with predictive validation of mechanistic IBMs applies to models in general: they are intentional simplifications of reality and therefore cannot and should not represent the myriad processes and events that affect real populations and communities. For example, when applied to a heavily managed and studied spawning stream in Northern California (Railsback et al. 2013), our Chinook salmon IBM's predictions of annual numbers of outmigrating juvenile salmon did not closely match data from a fish trap at the mouth of the stream. The poor relation between these primary results of the model and data was no doubt partly due to uncertainty in both the model and the trap data. However, in discussions of our results, we also learned about potentially highly influential events, such as forest fires in some years whose physical effects on the system had not been quantified. Our IBMs are complex and produce many testable results (e.g., the Chinook model could predict the size and outmigration time of juvenile salmon quite well). However, they are still models designed specifically to represent how fish populations respond to things like flow and temperature that we can at least partially manage. In this example, the model could have included the influence of forest fires on turbidity and water temperature had they been measured, but it is not designed to include all the effects of forest fires on aquatic ecosystems. Making the model more complex might make sense to address a specific new problem, but in general we face a trade-off between complexity and usability. Trying too hard to match data can result in models too complex to be useful.

The problems that often limit the potential to validate models by testing their predictions of annual variation in fundamental outputs, like inadequate information on variation in key drivers or too few years of data, can be fatal for validation of simple models. However, in the following sections we argue that these problems do not necessarily prevent mechanistic IBMs from gaining credibility and being extremely useful.

10.3 STRATEGIES FOR BUILDING CREDIBILITY

If validating an IBM by the traditional method of comparing its primary results to observed population data is impractical or impossible, then how can we assess, build, and communicate its credibility? We suggest two approaches: (a) embracing the concept of invalidation and (b) seeking to increase understanding of the model.

Holling (1978) considered model validation equivalent to hypothesis testing, and therefore focused on invalidation, because hypotheses can be rejected but not proved correct (Popper 1959): models gain credibility as they survive the risk of being proved wrong. The concept of invalidation forms the core of the pattern-oriented modeling (POM) approach that we explored in chapters 8 and 9. In POM, modelers select a set of observed patterns for use in the formulation, testing, and parameterization of IBMs. Each pattern presents an opportunity for model invalidation. POM also commonly incorporates another valuable element in assessing model credibility: the consideration of alternative models (Holling 1978, Augusiak et al. 2014). In the formulation of mechanistic models, consideration of alternative models often includes those that rely on different mechanisms and different numbers of mechanisms. Chapter 9 focuses on the specific case of contrasting alternative theories for adaptive behavior, but we can also use pattern-oriented invalidation experiments on full models.

The mechanistic nature of IBMs with SPT provides potential sources of invalidation via an array of patterns on both the individual and population levels of organization. These are available throughout the modeling cycle, from initial model formulation through what Augusiak et al. (2014) refer to as *model output corroboration*. Opportunities for invalidation of course include simulation experiments designed for that purpose, but also may arise haphazardly from new observations and new literature. Ideally, an IBM will gain credibility as the "weight of evidence" provided by the survival of invalidation opportunities accumulates.

The mechanistic nature of IBMs with SPT also provides unprecedented opportunities to observe and understand the workings of the models. As Holling (1978) states, "It is on understanding alone that a critical assessment of model credibility must ultimately be based." On many occasions, we have gained confidence in IBMs when initially surprising results made sense in retrospect. Of course the two approaches identified here are often linked: the process of attempting model invalidation almost inevitably adds to understanding, whether or not invalidation occurs.

10.4 LESSONS LEARNED IN FIELD, LABORATORY,
AND SIMULATION EXPERIMENTS

Here we list and briefly describe specific approaches that have proved useful to us and others for testing the credibility of IBMs with adaptive individual behavior. Regardless of the outcome in terms of model credibility, they commonly reveal research that could significantly improve our understanding of the modeled system. In fact, perhaps the most important lesson from the following discussion is that model credibility benefits when model analysis can be linked to empirical research in a cycle of testing and revision. As we explore in chapter 11, failures by model invalidation turn into research successes when they identify important knowledge gaps that can be resolved with empirical studies.

Bottom-up corroboration. It should be clear that the first requirement for credibility of mechanistic, "bottom-up" models such as IBMs is establishing the credibility of their separate mechanisms or submodels. The POM theory development cycle of chapter 9 is one part of bottom-up corroboration. In addition, the submodels for all other mechanisms (e.g., feeding, energetics, starvation, predation risk) should be thoroughly documented, explored, calibrated, and tested. Our model description documents are filled with graphs showing how each submodel behaves over the range of conditions it will encounter in the IBM.

We consider confirmation that individuals behave reasonably within the IBM a second essential component of bottom-up corroboration. Individual behavior should be evaluated in every simulation experiment; behavioral changes in response to treatments should make sense, either before or after the fact. For example, to make the inSTREAM trout model useful for predicting effects of forest management, which can affect erosion and therefore stream turbidity, we included the effects of turbidity on trout feeding success and predation risk. Before examining the population-level results of simulation experiments that contrasted different turbidity regimes, we confirmed that individual simulated fish responded reasonably to higher turbidity, for example, by occupying habitat with lower water velocity when high turbidity lowered the benefit of visual feeding on drifting prey (Harvey and Railsback 2009).

Pattern-oriented analysis of population responses. This approach involves testing how well the IBM reproduces observed patterns of population response to conditions relevant to the purpose of the model. The aforementioned simulation experiments on turbidity provide perhaps our most productive example of this approach. In that study, the comparison of model results to data revealed a conflict: the extinction of virtual populations, when we calibrated the model in the standard way, versus the persistence of natural populations facing the same turbidity regimes. This conflict suggested that the feeding submodel underestimated the ability of fish to acquire food under turbid conditions. That result then motivated field observations and laboratory experiments (White and Harvey 2007; Harvey and White 2008) that revealed substantial feeding success by trout in water too turbid for visual feeding, even though much of the extensive literature on trout feeding assumes trout depend on visual detection of prey drifting by in the water column (Harvey and Railsback 2014). Representing this nonvisual feeding ability in the model was relatively straightforward, because the original formulation of the model included both drift feeding and a second kind of feeding that did not depend on vision or water movement.

Pattern-oriented analysis of secondary predictions. In this approach, we compare model results other than primary population-level predictions to empirical knowledge. The empirical knowledge can simply be general patterns extracted from the literature. For example, an early version of inSTREAM was analyzed in a series of simulation experiments to determine the extent to which it produced general patterns, such as a "critical period" of high density-dependent mortality in juveniles, density-dependent growth, and lower abundance of large adults when pool habitat is reduced (Railsback et al. 2002). Even though not linked to a particular study site, these experiments have been very valuable as evidence that inSTREAM contains the mechanisms necessary to represent how population dynamics arise from interactions among individuals and their habitat.

Site-specific empirical data allow more detailed tests of model credibility. Model calibration requires field data describing key population-level characteristics, such as abundance and mean individual growth or some basis for reasonable approximation of those. But model results can then be tested against a variety of site-specific field data not directly related to the calibration process. For example, we have previously used the seasonality of growth, the distributions of individual growth, and the ability to predict results across study reaches following calibration to one of them (Harvey et al. 2014). We have also observed that good correspondence between predicted and actual habitat use at sites of interest strongly enhances the perceived credibility of IBMs in which habitat selection is the primary adaptive behavior.

This approach can be useful even in the absence of observations comparable to model results: showing that the model produces unobserved yet believable results lends credibility. For example, in a study that used inSTREAM to explore the watershed-scale effects of barriers to upstream movement (e.g., natural falls, dams, or poorly designed culverts) on population abundance and persistence (Harvey and Railsback 2012), we examined the life history characteristics of simulated subpopulations isolated above barriers. This effort revealed characteristics such as relatively high survival early in life and smaller size at reproduction that had been attributed by others to evolutionary processes not included in our model. The inSTREAM experiment therefore showed that adaptive behavior and its consequences can explain these life history patterns without evolution. The model has also produced various unanticipated patterns that were quite reasonable after we considered the effects of habitat on predation risk and food availability in combination with interactions among individuals. Models with reasonable representation of adaptive behavior at their core are likely to continue to produce unanticipated responses that make sense in retrospect, although at some point the frequency of these may say more about the limited vision of the modeler than the credibility of the model.

Sensitivity analysis. Of course modelers use sensitivity analyses to gain general understanding about a model, but reasonable results from sensitivity analyses also enhance credibility. In an IBM with adaptive trade-offs between growth and predation risk, sensitivity to parameters with clear links to survival and growth makes sense; sensitivity to parameters with narrow influence or those linked to processes unlikely to control population dynamics does not. IBMs as complex as inSTREAM must be expected to have a substantial number of parameters with small or modest effects, or strong effects only under limited conditions.

Sensitivity analyses also can motivate empirical research that contributes to model credibility or refinement. A complete sensitivity analysis for our original trout model (Cunningham 2007; summarized in Railsback et al. 2009) identified a parameter that describes the depth at which fish gain substantial protection from terrestrial predators as exceptionally influential. We did not predict this sensitivity, but it made sense in retrospect because the parameter directly affects predation rate and strongly influences habitat selection as virtual fish trade off predation risk and food acquisition: in small streams, the parameter essentially determines how much of the stream can be used by fish. This result led us to conduct field experiments to refine the little-studied relation between habitat depth and predation risk (Harvey and White 2017).

Comparison of calibrated parameter values to observations. IBMs incorporating SPT may offer another opportunity to test model credibility: empirical measurement of model calibration parameter values. This step may be particularly valuable for model credibility where calibration parameters drive adaptive behavior decisions and therefore link to critical mechanisms in the model. Such opportunities may be rare, however, because calibration parameters are often chosen in part because they are hard to measure or perhaps do not directly correspond with any real-world measurements. In our trout model, baseline food availability and predation risk strongly influence model results through both direct effects on vital rates and indirect effects from the adaptive behavior of individuals. We calibrate baseline values for food availability and predation risk by fitting the model to empirical observations of population abundance and individual size. We typically assume these two parameters are constant over space and time. This means that precise correspondence between calibration values and a set of specific empirical measurements would not be expected, but neither would severe dissimilarity. We confirmed the reasonableness of calibration values for food availability by making measurements during a field experiment that we later simulated in the trout model (Harvey and Railsback 2014). For predation risk, the baseline parameter in the trout model measures risk in habitat that offers the least relief from predation (i.e., shallow water exposed to bird predation and lacking hiding cover). We experimentally estimated risk in those habitats in study reaches where we had applied the model and observed correspondence with calibration values (Harvey and Nakamoto 2013).

10.5 CONCLUSIONS

Some individual-based population models containing adaptive behavior have been tested successfully using the predictive validation approach. The validation of their coastal bird model by Stillman et al. (2000) and Stillman and Goss-Custard (2010) is a prominent example, especially because the model was shown not only to predict a fundamental response to a major habitat alteration but also to make relatively accurate secondary predictions about the variables driving population response. However, in our experience good opportunities for classical validation of population models rarely arise. In any case, this form of validation is not sufficient for reaching conclusions about model credibility, particularly when modelers seek to make predictions about novel conditions. Fortunately, IBMs with mechanistic adaptive behavior at their core offer diverse opportunities to be placed at risk of invalidation and can gain credibility through survival of that process. These IBMs also offer unprecedented opportunities to explore and understand mechanisms, another key step in evaluating model credibility, particularly when applying models to novel conditions.

We have been lucky enough to combine, over many years, development and analysis of our salmonid IBMs with empirical research to test and improve their component submodels. These models were intended specifically for management applications, so credibility has always been a concern. A combination of simulation experiments, field studies, and laboratory experiments on individual fish have accumulated evidence that supports well-informed judgments of the models' credibility. We address the more general value of this combination in the next chapter.

CHAPTER 11

Empirical Research on Populations
of Adaptive Individuals

11.1 INTRODUCTION AND OBJECTIVES

One of the main reasons that our salmonid modeling efforts have been productive is that we conducted them as an integrated program of modeling and empirical research. From the start, we conducted field and laboratory studies on real animals to support the modeling effort. But it also became clear that modeling supports empirical research as well: the process of developing and testing the salmonid models identified a variety of critical yet under-studied questions.

The benefit of integrating modeling and empirical research has long been recognized (e.g., Platt 1964): theorists and modelers pose hypotheses that empirical researchers then design studies to test, and empirical research informs the development of new hypotheses. We believe such integration may be particularly valuable in frameworks that include multiple levels of organization, from individuals to populations to communities. But does working across levels of organization change the relationships of theory, modeling, and empirical research? What kinds of field and laboratory studies do we need, and at what levels of organization, to support modeling? And what do our across-level models tell us about new questions that empirical research can address?

We address these questions in this chapter. Thinking about the relation between modeling and empirical research requires us to address the entire process of model-based research, which is usefully characterized as a modeling cycle (figure 11.1; see also Grimm and Railsback 2005, Railsback and Grimm 2019). When initiating a research program, it is important to look at the full modeling cycle from the start: activities late in the cycle, such as model calibration and validation, often require the most extensive empirical data. This discussion relies on our own experience, but we also draw attention to the growing body of literature and experience from other programs that integrate individual-based modeling and empirical research. But before starting our discussion of empirical research through the modeling cycle in section 11.3, we first address how the kind of modeling and

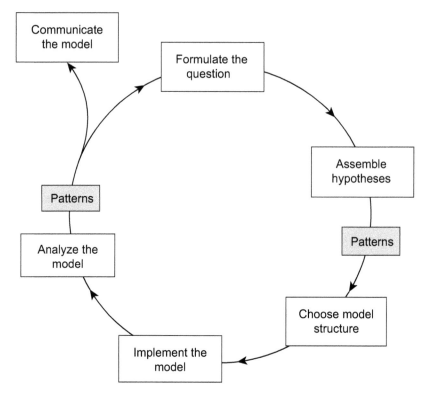

FIGURE 11.1. The modeling cycle (from Grimm and Railsback 2005). Modeling is depicted as tasks that start with formulating the question that the model addresses, and cycle through analyzing and communicating model results. Observed patterns inform the cycle by helping determine the model structure (section 8.2) and providing criteria for testing theory for adaptive behavior (section 9.2) and model validation (section 10.3).

theory development presented in this book can contribute to empirical studies and research.

Throughout this chapter we use the terms *model* and *modeling* for brevity, but we specifically refer to the individual-based, theory-driven population models discussed throughout this book.

11.2 BENEFITS OF MODELS FOR FIELD STUDIES

Models have traditionally received little use in addressing applied ecology questions, perhaps because traditional modeling approaches are too simple to address many of the problems ecologists work to solve. But this book is about IBMs designed to be useful for realistically complex problems. How might ecologists benefit from employing such models when addressing realistic management problems?

The most important benefit of integrating modeling with empirical research is that models can provide a framework for addressing questions too complex to be answered by field studies alone. Many research and management problems involve not only the complexities of adaptive behavior and individual variability but also interacting drivers (e.g., local habitat alteration and climate change) and spatiotemporal variation in conditions and processes. Attempts to address such problems using only field studies and statistical models often encounter basic study design problems, such as the inability to control and manipulate all the important drivers; the lack of adequate data before and after an impact; limited possible replication; and the absence of suitable control sites. Narrow, site-specific empirical studies also often provide little opportunity for general discoveries or contributions to theory.

When we combine field studies with mechanistic models, we can leverage the value of the empirical studies with the knowledge embedded in the model: instead of the field data being our only ammunition to attack the problem, we also have the theory and literature-derived information that goes into the model. Including mechanism and theory reduces the risks associated with predicting population dynamics beyond the range of observed conditions. The model also becomes a way of organizing the various kinds of knowledge we have, identifying the gaps to be filled, and making predictions with direct relevance to management decisions. Further, models can provide valuable frameworks for adaptive management programs: empirical studies can continue to test and refine the model, while the model can be used to design and evaluate management manipulations in a way likely to yield useful knowledge. The integration of modeling and empirical research also provides the opportunity to develop theory in the way we discussed in chapter 9 and, in general, to develop mechanistic understanding.

Modeling can also benefit field ecologists by providing a framework for interpretation of the kinds of data that we can now collect at massive rates. Many ecologists are using technologies such as GPS tracking, accelerometers, and environmental sensors to collect extensive databases on individuals and groups of individuals; yet translation of these data into useful knowledge, and especially into reliable theory for predicting individual behavior and its population consequences under novel conditions, remains a challenge (sections 1.3, 11.4). In fact, the trend toward extensive, unstructured data collection has been criticized. For example, Lindenmayer and Likens (2018) argue that progress in ecology and in protecting the environment requires not just data collection but also identification of important and tractable problems, formulation of conceptual models, and careful study designs. The modeling cycle we address in this chapter provides a framework for ensuring that data collection contributes to productive science: SPT and the theory development cycle provide a way to interpret patterns observed via sensor technology into general theory for use in population models. We discuss this potential further in section 11.4.

A third benefit of modeling is its ability to identify important research questions. Two examples from section 10.4 illustrate this point. In the first, sensitivity analysis of the IBM identified a process that was poorly understood yet very important to model results and population dynamics: the relation between water depth and predation risk. In the second example, the model predicted that trout could not persist in turbid conditions where in fact we do observe viable populations. This model failure inspired laboratory experiments that documented a mode of feeding that is generally ignored but sometimes critical to population persistence.

11.3 MODELING PHASE 1: FORMULATE THE QUESTION

The first phase of the modeling cycle carefully defines the problem to be addressed. Often a specific management problem determines the model's purpose; specification of a model's purpose can be more challenging when exploring basic ecology questions. Even when addressing management problems, however, questions are rarely specified completely and precisely enough to provide a clear guide to what needs to be in the model and, therefore, to identify especially useful empirical research. Development of a useful model often requires clarification of its purpose, ideally via collaboration between modelers and managers to document exactly what management questions the model needs to support.

Despite this common need to refine the question, our experience suggests that working on management problems has major advantages for formulating problems that give clear direction to both modeling and field research and, therefore, for making rapid progress and real discoveries. Starting with a vague, "theoretical" question (e.g., "emergence of home ranges in large carnivores"; "causes of diversity in temperate forests") makes it much harder to decide how to design and parameterize a model and to identify novel empirical studies likely to be useful.

11.4 MODELING PHASE 2: ASSEMBLE HYPOTHESES

The second phase of the modeling cycle addresses preliminary decisions about what needs to be in the model: its spatial and temporal scales, the kinds of individuals and what variables and behaviors they have, the habitat and environment variables, etc. For models using it, this phase includes the first step of SPT—defining the adaptive decision that it models. As we discussed in section 8.2, pattern-oriented modeling (POM) is a strategy for basing these decisions on an understanding of the real system and its individuals, especially observed patterns that seem to characterize behavior relevant to the question formulated in phase 1. For well-studied taxa and systems, this understanding and patterns for POM can be obtained primarily from the literature. For less-studied systems, field studies will be required. We discuss several kinds of empirical research likely to contribute to this phase: natural history studies, characteristic patterns, and use of sensor and tracking technologies.

Natural history studies may be necessary to develop adequate understanding of the individuals and system. While conventional census and demographic data can be useful in later phases, at this phase we are more interested in understanding the underlying processes that drive individual fitness and, therefore, demographics. Several basic questions to study at this point seem extremely obvious but are not always considered explicitly. First: What do the organisms eat and how do they feed? Second: What preys on the organisms, under what conditions, and how do the organisms avoid predation? Third: How do the organisms reproduce and what do they need for successful reproduction? Fourth: What are the organisms' most important adaptive behaviors and trade-off decisions relevant to the model question—how do they adapt to the kinds of changes under study to obtain resources, avoid predation, and reproduce?

Characteristic patterns useful for POM can be obtained from natural history studies. We discussed these patterns in section 8.2 (see also Grimm and Railsback 2005; Railsback and Grimm 2019), so here we only remind readers that useful patterns are less likely to be quantitative metrics (e.g., densities, demographic rates) and more likely to be qualitative trends and other responses to the changes the model is intended to address. To illustrate this point, we present examples of patterns that have been useful for designing several IBMs.

For the beech forest model of Rademacher et al. (2004; also described by Railsback and Grimm 2019), designed to guide design of old-growth forest reserves:

- A horizontal spatial mosaic of developmental stages from seedlings through closed, mature canopy, apparent at spatial resolutions around 15 m^2;
- High mortality of young trees under closed canopies; and
- Rapid growth of young trees in canopy gaps.

For the model of Railsback and Johnson (2011) of how land use and pest insect consumption by birds determine crop production on Jamaican coffee farms:

- In exclosures that eliminate birds, lower pest infestation rates in shade-grown coffee than in sun-grown coffee;

- Higher bird densities in shade-grown than sun-grown coffee;
- In exclosure experiments, the percentage reduction in pest infestation by birds increasing with the infestation rate;
- Birds rapidly finding, and preying on, short-term insect outbreaks;
- Changes in bird density over a season positively related to changes in insect density;
- Episodic pest consumption by birds; and
- The distance between locations of individual birds observed at the start and end of a 1-hour period following a log-normal distribution, with most birds appearing to move little but a few individuals moving long distances.

For a model of how river management affects breeding success of a frog that lays eggs in rivers (Railsback et al. 2016), the "patterns" used for model design were actually natural history observations:

- Breeding activity appears triggered by a threshold water temperature;
- Frogs delay egg-laying during fluctuating river flows;
- Eggs and tadpoles can be killed by increases in flow that wash them downstream and by decreases in flow that expose them to air; and
- Eggs develop at a rate driven mainly by temperature, while tadpole development appears controlled by a combination of factors that include temperature.

Sensor and tracking technologies have recently become affordable and popular and could, combined with appropriate methods for interpreting the data, be very useful at this modeling phase. Cameras mounted on animals can be used to directly observe feeding and other behaviors and perhaps provide information on predation risk. Accelerometers and other sensors can be used to estimate how much time animals spend in different behaviors or states that could possibly be directly related to the fitness measure used to model behavior. These studies often require experiments to relate the raw data to the behaviors or states of interest (Wilson et al. 2014).

The use of technologies that track animal locations over time is now the subject of an extensive literature that illuminates both the ease of data collection and the challenges of interpreting it. Our friend Justin Calabrese of the Smithsonian Conservation Biology Institute identifies five general objectives of tracking studies: (1) observing locations and evaluating movement metrics, such as speed and distance; (2) evaluating metrics of space use, such as home range size and occurrence areas (Noonan et al. 2019); (3) identifying behavioral states and when individuals change states (Gurarie et al. 2016); (4) relating location or movement to habitat variables, e.g., via resource selection and step selection functions (Thurfjell et al.

2014); and (5) understanding how individuals affect the movement of each other through social interactions (Calabrese et al. 2018).

All of these objectives are potentially relevant to this second phase of the modeling cycle. Models of adaptive behavior are already being informed by tracking data, if not in an explicitly pattern-oriented way; examples include models intended to explain adaptive foraging and migration behavior in beluga whales (Bailleul et al. 2013), ringed seals (Liukkonen et al. 2018), and harbor porpoises (Nabe-Nielsen et al. 2013). However, we remind readers that when modeling populations of individuals that exhibit adaptive behavior, at this phase we are less interested in reproducing observed locations and movement (which could be important in model calibration and validation; section 11.6) and more interested in *patterns* of how such measures *change* in response to the driving factors of concern in our model.

Ideally, before deploying sensors, ecologists will think about what kinds of patterns will reveal important processes and behaviors and design the field studies to observe them. Study design could also consider patterns that will be useful for developing and testing theory for adaptive behaviors, as we discuss in the next section. Often the same patterns can both inform model design and test theory for adaptive behavior.

11.5 MODELING PHASE 3: CHOOSE MODEL STRUCTURE

The third phase of the modeling cycle includes the detailed model design, including steps 2–4 of implementing SPT as outlined in chapters 3 and 8. This phase involves building detailed submodels for all the processes that link individual fitness and decision-making to the individual's internal state and to its environment. Therefore, the most useful empirical research at this phase is at the individual level, focusing on developing sufficient understanding of physiology and behavior. For many taxa, empirical research to support these submodels will be a priority, but the overall modeling cycle can proceed by using temporary, approximated submodels borrowed from other taxa.

Our experience has been almost entirely with vertebrate animals, so we can say little about what kind of research may be most commonly needed for plants, invertebrates, or other taxa. Models that use SPT or similar approaches to adaptive behavior typically need submodels for at least three kinds of processes: feeding, energetics, and risk.

Feeding submodels relate food intake and foraging energy costs to individual variables, such as size and life stage, and to environmental variables, such as food availability and habitat characteristics that affect feeding. Our trout model's feeding submodel relies on an extensive literature on how trout food intake and swimming speed depend on water velocity, depth, turbidity, temperature, fish length, and food density. In contrast, for the coffee farm model mentioned in section 11.4, we were forced to make up a simple submodel of how bird food intake depends on the density and "catchability" (ease of food detection and bird maneuvering) in each habitat type, and then identify this topic as a priority for empirical research (Railsback and Johnson 2014).

Energetics submodels represent how food intake and respiration costs determine the individual state variables directly related to fitness, such as size, energy reserves, starvation risk, and reproductive potential. Familiarity with SPT, DSVM, or any other fitness-based approaches to modeling behavior should make it clear that energetics is critical to behavioral and population ecology: energetics is what relates behavior—which determines energy intake and metabolic costs—to fitness. Models that include energetics as part of adaptive behavior are highly sensitive to the energetics submodel's parameters. Our trout model again benefited from an extensive literature that includes a standard energetics modeling approach for fish (Hanson et al. 1997). We anticipate, though, that as the importance of energetics to population ecology is better appreciated, there will be a strong demand for data and model development.

Even for taxa with extensive energetics data and models, we may need additional research to produce submodels adequate for modeling behavior with state-based approaches like SPT. When we assume that organisms make tradeoff decisions between, for example, predation risks and the risks resulting from inadequate energy intake, such as starvation, disease, and reduced reproductive potential, we must be able to model what happens to the organism's fitness as its energetic state declines. Understandably, but unfortunately, when physiological ecologists study and model energetics, they often avoid degraded energy states and their effects. We found, for example, that the extensive literature on trout energetics contains very little on growth or weight loss at low rations or stressfully high temperatures or on how risk of starvation and disease increases as energy stores are depleted. Particularly valuable research on starvation could also include efforts to both identify useful indicators of potential starvation risk and quantify the levels of those indicators that cause changes in behavior.

Predation and other risks are the third kind of submodel for which we anticipate a frequent need for empirical research. Understanding why individuals die and how risk varies with individual state and environmental conditions seems important for population ecology in general, and is especially important when risk is a major driver of behavior. But mortality is particularly difficult to observe and quantify, and therefore under-studied. Measurement of perceived risk (Brown 1988; Brown and Kotler 2004) avoids the problem of directly measuring mortality; and it is the perception of risk that actually drives behavior (figure 11.2).

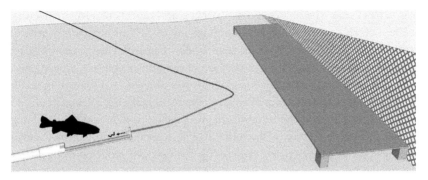

FIGURE 11.2. Experimental apparatus used by Harvey and White (2017) to quantify how predation risk perceived by trout varies with two habitat variables, water depth and distance to hiding cover. Perceived risk was quantified as "giving-up" food levels (Brown and Kotler 2004): the minimum rate of food availability that induces a fish to remain in a risky location. The feeding device (lower left) allows the experimenter to release food items that, if not captured by the fish immediately, are sucked away so that the fish must remain close to the device to feed successfully. Wild fish confined in large enclosures were trained to use the feeding device, and then it was moved across gradients of depth and distance from the platform (right) that provided hiding cover. At each location, the experimenter reduced the rate of food delivery gradually until the trout abandoned the feeder for a safer location. The experiment found, e.g., that giving-up food levels increase sharply with depth and no amount of food would induce fish to use depths < 20 cm. (Figure by Jason White.)

We end this discussion with a reminder that for population modeling we do not need, nor want, our IBM's submodels to contain the most detailed, accurate, or comprehensive understanding of processes such as feeding, energetics, or mortality risks. Instead, we need submodels that capture just enough of how these processes work to produce individual behavior and population dynamics adequate to address the model's purpose. This need for simplicity has two implications. First, we should not stop the modeling cycle and wait for empirical research when there is little information to base the submodel on. Instead, we can use a simple "placeholder" submodel and carry on. Second, when designing, funding, or collaborating on empirical research for these processes, we should keep in mind that the research need not be extremely detailed or innovative to be useful to a population model, and often we do not include everything we've learned from a field or lab study in the model. For our trout model, laboratory and field research to support submodels has focused on processes for which there is almost no literature (e.g., how predation risk varies with habitat characteristics) and on modeling energetics under rarely studied stressful, low-growth conditions.

The third phase of the modeling cycle can also include the POM theory development cycle that we discussed in chapter 9. Empirical research can have two roles in this cycle. The first is in hypothesizing new models of adaptive decision-making, which could include new formulations of SPT or completely different ways of modeling behavior. Behavioral studies can better define how the real organisms identify alternatives, predict future conditions, and select among alternatives; examples include understanding the use of cues, as discussed in section 8.4.1. The kinds of basic research on how organisms make decisions, discussed in section 1.4, will also, we hope, eventually become relevant to population ecology through this cycle. Certainly, the more we know about how organisms actually make decisions, the better we can specify the decision modeled using SPT or other methods.

The second role, we believe, is the most innovative and exciting link between modeling and empirical research: field studies specifically designed to identify patterns that can then be used to contrast and falsify alternative hypotheses—behavior submodels—in the POM theory development cycle. For our first two theory development exercises (described in sections 2.2 and 2.3), we were able to rely entirely on patterns already in the trout literature. For bird foraging theory used in the coffee farm model, we relied on patterns previously observed in field experiments in the same study system (Railsback and Johnson 2011). But the field study of Ainara Cortés-Avizanda (Cortés-Avizanda et al. 2010, 2016; discussed in section 9.3.1) is a particularly inspiring example because she designed a simple, elegant field study specifically to contrast, using POM, three competing theories for a vulture behavior. We hope the future of population ecology holds many more such combinations of modeling and field study to produce general, reusable theory for behaviors that explain population phenomena.

11.6 MODELING PHASE 5: ANALYZE THE MODEL

The final phase of the modeling cycle is to perform model analyses that typically include calibration, sensitivity and uncertainty analysis, validation, and simulation experiments to address the model's purpose. (The fourth phase, model implementation, does not involve empirical research.) For traditional population models built at the system level, the role of empirical studies at this phase is mainly to provide the demographic time series needed for calibration and validation (e.g., Hilborn and Mangel 1997). But across-level models such as IBMs with adaptive behavior offer many more opportunities for interaction between modeling and empirical research.

One key difference between analysis of the models we address in this book and traditional population models is that IBMs often represent much more detail about a population and therefore many more kinds of results to calibrate and validate.

Field studies for these purposes therefore benefit from collecting more information than simply census data; often models can be calibrated and tested much more thoroughly when the data include distributions of individual state variables (e.g., size, age) and how those distributions vary over space or other dimensions, such as habitat type. Also, many models are intended for purposes involving relations between populations and habitat and, therefore, require detailed physical (e.g., climatological, geographical, hydrological, or hydraulic) submodels and input to represent how habitat varies over time and space; modelers and field researchers must work together to collect such input as needed for model analyses.

Ideally, the interaction between modeling and empirical research during the model analysis phase will include not only models benefiting from input of empirical data but also identification of interesting and valuable empirical research opportunities. We discussed two such opportunities in section 11.2: using sensitivity and uncertainty analysis to identify submodels and parameters that are especially deserving of research, and using failed validation attempts to identify and investigate missing or misunderstood processes. Modeling and empirical research prove especially symbiotic when "problems" identified in model analysis are treated as research opportunities.

11.7 CONCLUSIONS

Chapter 1 discussed how traditional fields of ecology have so far made little progress toward useful general theory and models for populations and communities of adaptive individuals. This chapter shows that many different subdisciplines—examples of which include natural history, physiology, behavior, population and community ecology, and even neurobiology—are essential for tackling this problem with across-level models, such as IBMs that use SPT. When we model populations as collections of unique individuals with adaptive behavior, we need not just population data but empirical research and a variety of submodels designed to capture just enough of what individuals do to let us predict the dynamics of higher levels, and a variety of patterns observed at different levels and scales that contain information to design our models and test theories of adaptive behavior.

What kinds of field and laboratory studies do we need, and at what levels of organization, to support population modeling when we use SPT to represent adaptive behavior? Modelers still need population-level data, such as census time series for calibration and validation, but with more complex and realistic models we can use more than just population counts; we can test models against many more kinds of data. To represent behavior and its consequences, we also need individual-level submodels for a variety of physiological and behavioral mechanisms. But especially for lesser understood taxa, modeling must start with basic natural history knowledge, such as how individuals feed, what risks they must avoid, how they reproduce, and what their most important adaptive behaviors appear to be. And using POM to both design our IBM and test its theory for adaptive behavior requires observation of individual- and population-level patterns that characterize the mechanisms and behaviors that drive the population's dynamics.

What do our across-level models tell us about new questions that empirical research can address? Including the basic concepts of individual fitness in population models points to several research topics that seem under-studied for many taxa but of obvious importance. One is understanding and modeling the energetics that relate behavior to fitness: How do food intake and metabolic demands determine growth, reproductive potential, and survival of starvation? Even if we rarely observe individuals suffering from low energy reserves, we must understand the costs of reduced energy intake if we are to model the behaviors that trade off the benefits and costs of feeding. A second topic is more thorough understanding of predation and other mortality risks and how they vary with behavioral choices, such as when and where to forage. And for a surprising range of taxa we lack feeding models that predict the food intake and energy costs of foraging alternatives.

Empirical research is needed not just to address these questions for whatever taxa we choose to model but also to develop standardized submodels that can be used in IBMs for a variety of related taxa. Development of our trout model would have been much harder if not for a standard approach for modeling their feeding and a simple, standardized framework for the growth energetics of fish. Developing standard frameworks for modeling feeding, energetics, risk, etc., for guilds of similar organisms will be important.

Across-level models also change the way we study behavior. Much of the experimental literature on animal behavior has been conducted in highly simplified contexts that ignore both natural variability in environmental conditions and feedbacks from the behavior and resource use of other individuals. Our experience modeling bird foraging behavior in the coffee farm model, for example, showed that widely studied "classical" theory of foraging behavior was of little value in a population context (Railsback and Johnson 2011). For population models that include adaptive behavior, we need models of behavior that consider natural variability and uncertainty in environmental conditions, feedbacks from other individuals in the population or community, and reasonably realistic assumptions about decision-making mechanisms and abilities. These needs can be addressed both by studying individuals in population contexts and, from the bottom up, by studying the internal neurological and chemical mechanisms organisms use.

At the population level and larger scales of organization, another new challenge for empirical research is conducting field studies that do just a little more than censuses and movement tracking studies typically do. Ideally, field studies will be designed to quantify patterns needed for building IBMs and testing behavior theory—patterns driven by the mechanisms that need to be modeled. Modern technology makes it much easier to collect extensive datasets on individual and group behavior. These datasets become especially valuable for predicting population responses to novel conditions and future change when they are useful for building and testing mechanistic models.

CHAPTER 12

Conclusions and Outlook

12.1 MODELING POPULATIONS OF ADAPTIVE INDIVIDUALS

This book is a collaboration between an engineer and an aquatic ecologist, both of whom are interested in river management and why fish and their populations do what they do. Individual-based modeling provides an avenue to address these related goals.

We first tackled the problem of incorporating adaptive trade-off behavior in our IBMs: How can we model habitat selection in a way that realistically represents how fish trade off the benefits of growth against the risk of predation in a way that lets us predict population dynamics?

We soon perceived broad interest in the general question of how the adaptive trade-off behaviors of individuals affect population and community dynamics and, in turn, how those dynamics affect the individuals and their behavior. Terms and concepts such as *ecology of fear* (Brown et al. 1999) and *trait-mediated indirect effects* (Abrams 1995) arose to describe how adaptive trade-off behaviors affect population and community ecology. Since Peter Abrams's seminal paper showing how adaptive behavior invalidates fundamental assumptions of the Lotka-Volterra equations (Abrams 1993), we have known that we need new ways to model population and community dynamics that account for behavior.

As ecologists have developed a broad empirical understanding of trade-off behaviors and their effects, theory and modeling has fallen behind. Representing behavior in traditional differential equation models has not proved viable, especially when individual variation in behavior cannot be ignored. Individual-based modeling seems like a natural approach, but individual-based modelers have not yet adopted a common, general, and proved body of theory for adaptive trade-off behaviors in IBMs. Many modelers are attracted to optimization approaches such as DSVM but then realize that optimization is not feasible or realistic in many population models. Perhaps due to ecology's strong roots in mathematics, replacing optimization with good enough approximation has not been the natural response of many modelers. Our 2013 review (Railsback and Harvey 2013) found no theoretical approaches other than SPT that have caused an IBM to produce even the basic nonconsumptive and indirect effects of the simple tri-trophic system described in section 1.1.

Our goal in this book was to lay out one approach for modeling, studying, and understanding populations and communities of adaptive individuals. The approach includes these components:

- Individual-based models carefully designed to address specific problems of real populations and environments.
- Theory for how the IBMs individuals make adaptive trade-off decisions in their specific contexts. SPT provides a general framework for modeling behavior based on the fundamental concept of individual fitness. SPT uses approximation and explicit prediction to make good decisions when optimization is neither feasible nor realistic.
- Hypothesizing and contrasting alternative theories for adaptive behavior, and testing whether IBMs represent behavior with sufficient realism to make them useful for the population-level problems they were designed for.
- Linking modeling with empirical research in a cycle that refines IBMs and identifies research topics that contribute more directly to understanding interactions between the individual and higher levels of ecology.

12.2 KEY CHARACTERISTICS OF THE APPROACH

Key characteristics of our approach are noted throughout this book; we summarize them here. Many of these characteristics are shared with individual-based population modeling in general.

First, SPT explicitly includes multiple levels of organization: models of *individual* adaptive behavior explain *population* and higher-level dynamics. Traditionally, adaptive trade-off decisions have been the domain of behavioral ecologists working at the individual level and using optimization methods such as DSVM as their framework for thinking about and modeling decisions. SPT provides a framework for thinking about and modeling populations driven by trade-off behaviors, at subevolutionary time scales. Simply implementing the basic concepts of DSVM without the mathematical optimization—constraining our model individuals in the same way real organisms may be constrained, so they must rely on inaccurate predictions and approximations—is our "trick" for doing so.

Second, we find doing theory more productive when we model real systems and applied problems. Traditional theoretical ecology that relies on conventional mathematics often cannot address the messy problems management ecologists deal with every day. Alternatively, simulation modeling, when applied to messy, real problems, has usually been theory-free and hence of questionable general value (Grimm 1999). In contrast, our experience has been that theory development is much easier and more productive when we work on specific real systems from which we can extract the patterns that help us design models, identify the behaviors that need to be in the models, and test theory for those behaviors. SPT, like DSVM, demands submodels representing fitness-related processes, such as predation risk, growth and starvation risk, and reproductive energetics; it is much easier to find useful models and parameters for such processes when we work with real systems. When we think of developing models of adaptive behavior for use in IBMs as "doing theory," applied and theoretical ecology unite.

Third, as necessitated by the problems we face, the approach relies on simulation, approximation, and updating instead of traditional mathematical frameworks, such as differential equations, optimization, and game theory. Real organisms, like those in our IBMs, face decisions complex enough that optimization is simply not feasible. So while our models may be less mathematically satisfying, they may be more realistic as well as more useful for understanding real populations. This movement away from fitness optimization is in some ways analogous to arguments against the "adaptationist programme" that assumes evolution has optimized individual traits. For example, Gould and Lewontin (1979) advocate a view of evolution that acknowledges that traits can be so limited by trade-offs, interactions, and constraints that they cannot be completely optimal, and therefore that understanding such limitations can be just as important as understanding the optimizing forces. This view certainly seems applicable to the adaptive trade-off behaviors we address.

Fourth, SPT and the theory testing cycle facilitate beneficial linkage of modeling and empirical work. Formulation of models using SPT demands knowledge of the real organisms and systems being modeled, in part to represent the physiological and cognitive processes driving behavior. Testing of theory and models also relies on empirical knowledge. And when the models fail to reproduce observations, we learn what aspects of the real system we need to know more about. Using models to identify important unanswered research questions ensures that field and laboratory studies will contribute to improving models that advance ecological management.

Finally, the pattern-oriented theory development cycle gives ecologists the power of *strong inference*. Many studies, especially in behavioral ecology, have "tested" theory by examining whether data collected under simplified conditions correspond with the theory. But far fewer have hypothesized competing theories and contrasted them by their ability to reproduce multiple, diverse patterns observed in real, unmanipulated systems. We consider this way of producing ecological theory exciting, potentially productive, and largely unexplored.

12.3 CONCLUSIONS FROM EXAMPLE MODELS

In chapters 4 through 7 we examined a series of increasingly complex applications of SPT, in most cases exploring several alternative versions of the adaptive trade-off theory. Here we summarize lessons from those examples that we think are of general value for potential users of SPT.

The first lesson is that simple and wrong assumptions can produce good, even near optimal, behavior. Assumptions as simplistic as predicting that current conditions will persist until a distant time horizon, while clearly wrong, can give model individuals highly adaptive trade-off behaviors. The idea that unrealistic assumptions can still produce useful behavior in models has been discussed and accepted in other fields (e.g., in economics by Lehtinen and Kuorikoski 2007).

Second, the use of such simple assumptions and approximations can make it easy to include more realistic "details," such as continuous instead of binary relations between energy reserves and starvation risk. Such details can make behavior and population dynamics much more realistic and interesting, providing a high payoff for the cost of giving up strict optimization.

Third, in contexts lacking a natural time horizon, such as a life history transition, "sliding" and "leaping" time horizons can avoid undesirable artifacts of fixed time horizons.

Fourth, contingent decisions, such as habitat selection and choice among life history options that depend on outcomes of habitat selection, can be relatively easy to represent in SPT (as they are in DSVM).

Fifth, using equations for prediction (e.g., equation 8-2) is beneficial when feasible, but model individuals can also use simulation as a way to predict the future consequences of decision alternatives. The computational costs of simulation are not necessarily prohibitive.

However, we observed that more detailed and realistic prediction methods do not always produce important improvements in simulated behavior or population predictions. Also, greater detail can mean greater challenges in calibrating and understanding models. Detail should be added stepwise while testing its effects.

The preceding lessons suggest SPT can have advantages over DSVM and other optimization-based approaches for modeling behavior and life history decisions, even in models of single individuals. In addition, individual-level SPT models can usually be easily elaborated into individual-based population (or community) models that include complexities such as individual variation, interactions among individuals, feedbacks of behavior, and spatial and temporal variation in habitat.

Finally, we learned from the facultative anadromy model that modeling *popu-lations* of adaptive individuals can lead to quite different conclusions than does examining the optimal behavior of a single representative *individual*. The differ-ences often result from two factors: mortality and individual variation. Individual-level models tend to focus on the indirect effects of mortality risk via trade-off behaviors, but do not drive home that the direct effects of risk on abundance can also be strong. Variation among individuals in internal state and habitat can quickly increase due to the feedbacks of behavior and competition (the strong get stronger, etc.), so modeling only an average or typical individual can miss other individuals that may be very important to populations.

12.4 OUTLOOK

We hope that SPT and the pattern-oriented theory development cycle will foster advances in both ecological modeling and ecological research. Ideally, advances in modeling will include development of general theory that links individual behavior and population dynamics. Here we refer to "general theory" in the way that the physical sciences have: as reusable models that have proved useful for solving problems about the real world. We anticipate general theory develop-ing as people tackle ecological management problems, figure out what adaptive behaviors need to be modeled, and find theory that works in each case. Accu-mulated experience will eventually expose general, reusable theory: fitness mea-sures, prediction assumptions, and decision algorithms that are each shown useful in particular contexts. As suggested in chapter 11, we believe that seeking pat-terns useful for testing and contrasting alternative theory for adaptive behavior can make important contributions to population and community ecology and that development of IBMs that incorporate adaptive behavior can identify particularly valuable empirical research.

The research program we describe here—collecting empirical observations of a system and its individuals, building an IBM, and developing theory for adaptive trade-off behaviors that explain system dynamics—would be challenging for any one investigator. The program could require multiple field or laboratory experi-ments that may individually or collectively cover multiple ecological levels of organization and spatiotemporal scales, while building IBMs requires software development, which few ecologists are trained in. Further, the whole approach is relatively new, so relevant examples, guidance, and competent reviewers may be hard to find. But these challenges seem to be lessening. Availability of software such as NetLogo and instructional materials has dramatically reduced the software challenges. Individual-based approaches are gaining acceptance as a credible and necessary way to study ecology, and our experience has been that acceptance is more likely when an IBM's representation of adaptive behavior has been tested in the way we discuss in chapter 9. Cohesive research programs well justify and often enhance the value of separate studies. It appears that all the pieces are in place for ecologists to make rapid progress building and testing useful models of populations and communities of adaptive individuals.

References

Abrams, P. A. 1993. Why predation rate should not be proportional to predator density. *Ecology* 74: 726–33.

———. 1995. Implications of dynamically variable traits for identifying, classifying, and measuring direct and indirect effects in ecological communities. *American Naturalist* 146: 112–34.

———. 2007. Defining and measuring the impact of dynamic traits on interspecific interactions. *Ecology* 88: 2555–62.

———. 2010. Implications of flexible foraging for interspecific interactions: Lessons from simple models. *Functional Ecology* 24: 7–17.

Alonzo, S. H. 2002. State-dependent habitat selection games between predators and prey: The importance of behavioral interactions and expected lifetime reproductive success. *Evolutionary Ecology Research* 4: 759–78.

Alonzo, S. H., P. V. Switzer, and M. Mangel. 2003. Ecological games in space and time: The distribution and abundance of Antarctic krill and penguins. *Ecology* 84: 1598–607.

Amano, T., K. Ushiyama, S. Moriguchi, G. Fujita, and H. Higuchi. 2006. Decision-making in group foragers with incomplete information: Test of individual-based model in geese. *Ecological Monographs* 76: 601–16.

Antonsson, T., and S. Gudjonsson. 2002. Variability in timing and characteristics of Atlantic salmon smolt in Icelandic rivers. *Transactions of the American Fisheries Society* 131: 643–55.

Aoki, I. 1982. A simulation study on the schooling mechanism in fish. *Bulletin of the Japanese Society of Scientific Fisheries* 48: 1081–88.

Audzijonyte, A., and S. A. Richards. 2018. The energetic cost of reproduction and its effect on optimal life-history strategies. *American Naturalist* 192: E150-62.

Augusiak, J., P. J. Van den Brink, and V. Grimm. 2014. Merging validation and evaluation of ecological models to 'evaludation': A review of terminology and a practical approach. *Ecological Modelling* 280: 117–28.

Auyang, S. Y. 1998. *Foundations of Complex-System Theories in Economics, Evolutionary Biology, and Statistical Physics*. Cambridge University Press, New York.

Bailey, D. W., J. E. Gross, E. A. Laca, L. R. Rittenhouse, M. B. Coughenour, D. M. Swift, and P. L. Sims. 1996. Mechanisms that result in large herbivore grazing distribution patterns. *Journal of Range Management* 49: 386–400.

Bailleul, F., V. Grimm, C. Chion, and M. Hammill. 2013. Modeling implications of food resource aggregation on animal migration phenology. *Ecology and Evolution* 3: 2535–46.

Ballerini, M., N. Cabibbo, R. Candelier, A. Cavagna, E. Cisbani, I. Giardina, V. Lecomte, A. Orlandi, G. Parisi, A. Procaccini, M. Viale, and V. Zdravkovic. 2008. Interaction ruling animal collective behavior depends on topological rather than metric distance: Evidence from a field study. *Proceedings of the National Academy of Sciences* 105: 1232–37.

Barrett, J. C., G. D. Grossman, and J. Rosenfeld. 1992. Turbidity-induced changes in reactive distance of rainbow trout. *Transactions of the American Fisheries Society* 121: 437–43.

Beckerman, A., O. L. Petchey, and P. J. Morin. 2010. Adaptive foragers and community ecology: Linking individuals to communities and ecosystems. *Functional Ecology* 24: 1–6.

Bergeron, P., D. Réale, M. M. Humphries, and D. Garant. 2011. Anticipation and tracking of pulsed resources drive population dynamics in eastern chipmunks. *Ecology* 92: 2027–34.

Bonnell, T. R., M. Campennì, C. A. Chapman, J. F. Gogarten, R. A. Reyna-Hurtado, J. A. Teichroeb, M. D. Wasserman, and R. Sengupta. 2013. Emergent group level navigation: An agent-based evaluation of movement patterns in a folivorous primate. *PLOS ONE* 8: e78264.

Boutin, S., L. A. Wauters, A. G. McAdam, M. M. Humphries, G. Tosi, and A. A. Dhondt. 2006. Anticipatory reproduction and population growth in seed predators. *Science* 314: 1928–30.

Brown, J. S. 1988. Patch use as an indicator of habitat preference, predation risk, and competition. *Behavioral Ecology and Sociobiology* 22: 37–47.

Brown, J. S. 2016. Why Darwin would have loved evolutionary game theory. *Proceedings of the Royal Society B* 283: 20160847.

Brown, J. S., and B. P. Kotler. 2004. Hazardous duty pay and the foraging cost of predation. *Ecology Letters* 7: 999–1014.

Brown, J. S., J. W. Laundré, and M. Gurung. 1999. The ecology of fear: Optimal foraging, game theory and trophic interactions. *Journal of Mammalogy* 80: 385–99.

Budaev, S., J. Giske, and S. Eliassen. 2018. AHA: A general cognitive architecture for Darwinian agents. *Biologically Inspired Cognitive Architectures* 25: 51–57.

Calabrese, J. M., C. H. Fleming, W. F. Fagan, M. Rimmler, P. Kaczensky, S. Bewick, P. Leimgruber, and T. Mueller. 2018. Disentangling social interactions and environmental drivers in multi-individual wildlife tracking data. *Philosophical Transactions of the Royal Society B: Biological Sciences* 373: 20170007.

Caro, T. 2007. Behavior and conservation: A bridge too far? *Trends in Ecology and Evolution* 22: 394–400.

Charnov, E. L. 1976. Optimal foraging, the marginal value theorem. *Theoretical Population Biology* 9: 129–36.

Clark, A. 2013. Whatever next? Predictive brains, situated agents, and the future of cognitive science. *Behavioral and Brain Sciences* 36: 181–204.

Clark, C. W., and M. Mangel. 2000. *Dynamic State Variable Models in Ecology*. Oxford University Press, New York.

Clark, M. E., and K. A. Rose. 1997. Individual-based model of stream-resident rainbow trout and brook char: Model description, corroboration, and effects of sympatry and spawning season duration. *Ecological Modelling* 94: 157–75.

Cook, R. C., D. L. Murray, J. G. Cook, P. Zager, and S. L. Monfort. 2001. Nutritional influences on breeding dynamics in elk. *Canadian Journal of Zoology* 79: 845–53.

Cortés-Avizanda, A., G. Blanco, T. L. DeVault, A. Markandya, M. Z. Virani, J. Brandt, and J. A. Donázar. 2016. Supplementary feeding and endangered avian scavengers:

Benefits, caveats, and controversies. *Frontiers in Ecology and the Environment* 14: 191–99.

Cortés-Avizanda, A., M. Carrete, and J. A. Donázar. 2010. Managing supplementary feeding for avian scavengers: Guidelines for optimal design using ecological criteria. *Biological Conservation* 143: 1707–15.

Cortés-Avizanda, A., R. Jovani, J. A. Donázar, and V. Grimm. 2014. Bird sky networks: How do avian scavengers use social information to find carrion? *Ecology* 95: 1799–808.

Courter, I. I., D. B. Child, J. A. Hobbs, T. M. Garrison, J. J. G. Glessner, and S. Duery. 2013. Resident rainbow trout produce anadromous offspring in a large interior watershed. *Canadian Journal of Fisheries and Aquatic Sciences* 70: 701–10.

Cunningham, P. M. 2007. A sensitivity analysis of an individual-based trout model. MS thesis, Humboldt State University, Arcata, CA.

Dill, L. M. 2017. Behavioural ecology and marine conservation: A bridge over troubled water? *ICES Journal of Marine Science* 74: 1514–21.

Eliassen, S., B. S. Andersen, C. Jørgensen, and J. Giske. 2016. From sensing to emergent adaptations: Modelling the proximate architecture for decision-making. *Ecological Modelling* 326: 90–100.

Emlen, J. M. 1966. The role of time and energy in food preference. *American Naturalist* 100: 611–17.

Enright, J. 1977. Diurnal vertical migration: Adaptive significance and timing. Part 1. Selective advantage: A metabolic model. *Limnology and Oceanography* 22: 856–72.

Fiksen, Ø. 1997. Allocation patterns and diel vertical migration: Modeling the optimal *Daphnia*. *Ecology* 78: 1446–56.

Forbes, V. E., S. Railsback, C. Accolla, B. Birnir, R. J. F. Bruins, V. Ducrot, N. Galic, K. Garber, B. C. Harvey, H. I. Jager, A. Kanarek, R. Pastorok, et al. 2019. Predicting impacts of chemicals from organisms to ecosystem service delivery: A case study of endocrine disruptor effects on trout. *Science of the Total Environment* 649: 949–59.

Foster, M. S. 1987. Delayed maturation, neoteny, and social system differences in two manakins of the genus *Chiroxiphia*. *Evolution* 41: 547–58.

Geritz, S. A., G. Mesze, and J. A. Metz. 1998. Evolutionarily singular strategies and the adaptive growth and branching of the evolutionary tree. *Evolutionary Ecology* 12: 35–57.

Gigerenzer, G., and R. Selten. 2002. *Bounded Rationality: The Adaptive Toolbox.* MIT Press, Cambridge, MA.

Gigerenzer, G., P. M. Todd, and A. R. Group. 1999. *Simple Heuristics That Make Us Smart.* Oxford University Press, New York.

Gilliam, J. F., and D. F. Fraser. 1987. Habitat selection under predation hazard: Test of a model with foraging minnows. *Ecology* 68: 1856–62.

Giske, J., S. Eliassen, Ø. Fiksen, P. J. Jakobsen, D. L. Aksnes, C. Jørgensen, and M. Mangel. 2013. Effects of the emotion system on adaptive behavior. *American Naturalist* 182: 689–703.

Giske, J., S. Eliassen, Ø. Fiksen, P. J. Jakobsen, D. L. Aksnes, M. Mangel, and C. Jørgensen. 2014. The emotion system promotes diversity and evolvability. *Proceedings of the Royal Society B* 281: 20141096.

Glimcher, P. W. 2016. Proximate mechanisms of individual decision-making behavior. Pages 85–96 *in* D. S. Wilson and A. Kirman, editors, *Complexity and Evolution: Toward a New Synthesis for Economics.* MIT Press, Cambridge, MA.

Gliwicz, M. Z. 1986. Predation and the evolution of vertical migration in zooplankton. *Nature* 320: 746.

Gould, S. J., and R. C. Lewontin. 1979. The spandrels of San Marco and the Panglossian paradigm: A critique of the adaptationist programme. *Proceedings of the Royal Society B* 205: 581–98.

Gray, D. R., and G. A. Williams. 2010. Knowing when to stop: Rhythms of locomotor activity in the high-shore limpet, *Cellana grata* Gould. *Journal of Experimental Marine Biology and Ecology* 391: 125–30.

Grimm, V. 1999. Ten years of individual-based modelling in ecology: What have we learned and what could we learn in the future? *Ecological Modelling* 115: 129–48.

Grimm, V., D. Ayllón, and S. F. Railsback. 2017. Next-generation individual-based models integrate biodiversity and ecosystems: Yes we can, and yes we must. *Ecosystems* 20: 229–36.

Grimm, V., U. Berger, D. L. DeAngelis, G. Polhill, J. Giske, and S. F. Railsback. 2010. The ODD protocol: A review and first update. *Ecological Modelling* 221: 2760–68.

Grimm, V., and S. F. Railsback. 2005. *Individual-Based Modeling and Ecology*. Princeton University Press, Princeton, NJ.

———. 2012. Pattern-oriented modelling: A "multiscope" for predictive systems ecology. *Philosophical Transactions of the Royal Society B* 367: 298–310.

Grimm, V., E. Revilla, U. Berger, F. Jeltsch, W. M. Mooij, S. F. Railsback, H.-H. Thulke, J. Weiner, T. Wiegand, and D. L. DeAngelis. 2005. Pattern-oriented modeling of agent-based complex systems: Lessons from ecology. *Science* 310: 987–91.

Gross, M. R., R. M. Coleman, and R. M. McDowell. 1988. Aquatic productivity and the evolution of diadromous fish migration. *Science* 239: 1291–93.

Gurarie, E., C. Bracis, M. Delgado, T. D. Meckley, I. Kojola, and C. M. Wagner. 2016. What is the animal doing? Tools for exploring behavioural structure in animal movements. *Journal of Animal Ecology* 85: 69–84.

Hanson, P., T. Johnson, J. Kitchell, and D. E. Schindler. 1997. *Fish Bioenergetics 3.0*. University of Wisconsin Sea Grant Institute, Madison, WI.

Harvey, B. C. 1991. Interactions among stream fishes: Predator-induced habitat shifts and larval survival. *Oecologia* 87: 29–36.

Harvey, B. C., and R. J. Nakamoto. 1999. Diel and seasonal movements by adult Sacramento pikeminnow (*Ptychocheilus grandis*) in the Eel River, northwestern California. *Ecology of Freshwater Fish* 8: 209–15.

———. 2013. Seasonal and among-stream variation in predator encounter rates for fish prey. *Transactions of the American Fisheries Society* 142: 621–27.

Harvey, B. C., R. J. Nakamoto, and J. L. White. 1999. Influence of large woody debris and a bankfull flood on movement of adult resident coastal cutthroat trout (*Oncorhynchus clarki*) during fall and winter. *Canadian Journal of Fisheries and Aquatic Sciences* 56: 2161–66.

Harvey, B. C., R. J. Nakamoto, J. L. White, and S. F. Railsback. 2014. Effects of streamflow diversion on a fish population: Combining empirical data and individual-based models in a site-specific evaluation. *North American Journal of Fisheries Management* 34: 247–57.

Harvey, B. C., and S. F. Railsback. 2007. Estimating multi-factor cumulative watershed effects on fish populations with an individual-based model. *Fisheries* 32: 292–98.

———. 2009. Exploring the persistence of stream-dwelling trout populations under alternative real-world turbidity regimes with an individual-based model. *Transactions of the American Fisheries Society* 138: 348–60.

———. 2012. Effects of passage barriers on demographics and stability properties of a virtual trout population. *River Research and Applications* 28: 479–89.

———. 2014. Feeding modes in stream salmonid population models: Is drift feeding the whole story? *Environmental Biology of Fishes* 97: 615–25.

Harvey, B. C., and J. L. White. 2008. Use of benthic prey by salmonids under turbid conditions in a laboratory stream. *Transactions of the American Fisheries Society* 137: 1756–63.

———. 2016. Use of cover for concealment behavior by rainbow trout: Influences of cover structure and area. *North American Journal of Fisheries Management* 36: 1308–14.

———. 2017. Axes of fear for stream fish: Water depth and distance to cover. *Environmental Biology of Fishes* 100: 565–73.

Hemelrijk, C. K., and H. Hildenbrandt. 2015. Diffusion and topological neighbours in flocks of starlings: Relating a model to empirical data. *PLOS ONE* 10: e0126913.

Hilborn, R., and M. Mangel. 1997. *The Ecological Detective: Confronting Models with Data*. Princeton University Press, Princeton, NJ.

Hill, J., and G. D. Grossman. 1993. An energetic model of microhabitat use for rainbow trout and rosyside dace. *Ecology* 74: 685–98.

Hodge, B. W., M. A. Wilzbach, W. G. Duffy, R. M. Quiñones, and J. A. Hobbs. 2016. Life history diversity in Klamath River steelhead. *Transactions of the American Fisheries Society* 145: 227–38.

Holling, C. S., editor. 1978. *Adaptive Environmental Assessment and Management*. John Wiley & Sons, Chichester, UK.

Houston, A. I., and J. M. McNamara. 1999. *Models of Adaptive Behaviour: An Approach Based on State*. Cambridge University Press, Cambridge, UK.

Hughes, N. F. 1992. Selection of positions by drift-feeding salmonids in dominance hierarchies: Model and test for arctic grayling (*Thymallus arcticus*) in subarctic mountain streams, interior Alaska. *Canadian Journal of Fisheries and Aquatic Sciences* 49: 1999–2008.

Hughes, N. F., and L. M. Dill. 1990. Position choice by drift-feeding salmonids: Model and test for arctic grayling (*Thymallus arcticus*) in subarctic mountain streams, interior Alaska. *Canadian Journal of Fisheries and Aquatic Sciences* 47: 2039–48.

Huse, G., and J. Giske. 1998. Ecology in Mare Pentium: An individual-based model for fish with adapted behavior. *Fisheries Research* 37: 163–78.

Huse, G., E. Strand, and J. Giske. 1999. Implementing behavior in individual-based models using artificial neural networks and genetic algorithms. *Evolutionary Ecology* 13: 469–83.

Huth, A., and C. Wissel. 1992. The simulation of the movement of fish schools. *Journal of Theoretical Biology* 156: 365–85.

Jager, H. I., D. L. DeAngelis, M. J. Sale, W. V. Van Winkle, D. D. Schmoyer, M. J. Sabo, D. J. Orth, and J. A. Lukas. 1993. An individual-based model for smallmouth bass reproduction and young-of-year dynamics in streams. *Rivers* 4: 91–113.

Kanno, Y., J. C. Vokoun, and B. H. Letcher. 2011. Sibship reconstruction for inferring mating systems, dispersal and effective population size in headwater brook trout (*Salvelinus fontinalis*) populations. *Conservation Genetics* 12: 619–28.

Kooijman, S. A. L. M. 1993. *Dynamic Energy Budgets in Biological Systems*. Cambridge University Press, Cambridge, UK.

———. 2010. *Dynamic Energy Budget Theory for Metabolic Organisation*. Cambridge University Press, Cambridge, UK.

Kułakowska, K. A., T. M. Kułakowski, I. R. Inglis, G. C. Smith, P. J. Haynes, P. Prosser, P. Thorbek, and R. M. Sibly. 2014. Using an individual-based model to select among alternative foraging strategies of woodpigeons: Data support a memory-based model with a flocking mechanism. *Ecological Modelling* 280: 89–101.

Lehtinen, A., and J. Kuorikoski. 2007. Unrealistic assumptions in rational choice theory. *Philosophy of the Social Sciences* 37: 115–38.

Lima, S. L., and L. M. Dill. 1990. Behavioral decisions made under the risk of predation: A review and prospectus. *Canadian Journal of Zoology* 68: 619–40.

Lindenmayer, D. F., and G. E. Likens. 2018. Maintaining the culture of ecology. *Frontiers in Ecology and the Environment* 16: 195.

Liukkonen, L., D. Ayllón, M. Kunnasranta, M. Niemi, J. Nabe-Nielsen, V. Grimm, and A.-M. Nyman. 2018. Modelling movements of Saimaa ringed seals using an individual-based approach. *Ecological Modelling* 368: 321–35.

Loeuille, N. 2010. Consequences of adaptive foraging in diverse communities. *Functional Ecology* 24: 18–27.

Loose, C., and P. Dawidowicz. 1994. Trade-offs in diel vertical migration by zooplankton: The costs of predator avoidance. *Ecology* 78: 2255–63.

Luttbeg, B., L. Rowe, and M. Mangel. 2003. Prey state and experimental design affect relative size of trait- and density-mediated indirect effects. *Ecology* 84: 1140–50.

Luttbeg, B., and O. J. Schmitz. 2000. Predator and prey models with flexible individual behavior and imperfect information. *American Naturalist* 155: 669–83.

Luttbeg, B., and G. C. Trussell. 2013. How the informational environment shapes how prey estimate predation risk and the resulting indirect effects of predators. *American Naturalist* 181: 182–94.

MacArthur, R. H., and E. R. Pianka. 1966. On optimal use of a patchy environment. *American Naturalist* 100: 603–9.

Mangel, M. 1994. Life history variation and salmonid conservation. *Conservation Biology* 8: 879–80.

Mangel, M., and C. W. Clark. 1986. Toward a unified foraging theory. *Ecology* 67: 1127–38.

———. 1988. *Dynamic Modeling in Behavioral Ecology*. Princeton University Press, Princeton, NJ.

Mangel, M., and W. H. Satterthwaite. 2008. Combining proximate and ultimate approaches to understand life history variation in salmonids with application to fisheries, conservation, and aquaculture. *Bulletin of Marine Science* 83: 107–30.

Manly, B. F. 1997. *Randomization, Bootstrap and Monte Carlo Methods in Biology*. 2nd ed. Chapman and Hall/CRC, Boca Raton, FL.

McLaren, I. A. 1963. Effects of temperature on growth of zooplankton, and the adaptive value of vertical migration. *Journal of the Fisheries Board of Canada* 20: 685–727.

McNamara, J. M., R. F. Green, and O. Olsson. 2006. Bayes' theorem and its applications in animal behaviour. *Oikos* 112: 243–51.

Metcalfe, N. B., N. H. C. Fraser, and M. D. Burns. 1998. State-dependent shifts between nocturnal and diurnal activity in salmon. *Proceedings of the Royal Society B* 265: 1503–7.

————. 1999. Food availability and the nocturnal vs. diurnal foraging trade-off in juvenile salmon. *Journal of Animal Ecology* 68: 371–81.

Miner, B. G., S. E. Sultan, S. G. Morgan, D. K. Padilla, and R. A. Relyea. 2005. Ecological consequences of phenotypic plasticity. *Trends in Ecology & Evolution* 20: 685–92.

Mitchell, M., and C. E. Taylor. 1999. Evolutionary computation: An overview. *Annual Review of Ecology and Systematics* 20: 593–616.

Morales, J. M., D. Fortin, J. L. Frair, and E. H. Merrill. 2005. Adaptive models for large herbivore movements in heterogeneous landscapes. *Landscape Ecology* 20: 301–16.

Mumme, R. L. 2018. The trade-off between molt and parental care in hooded warblers: Simultaneous rectrix molt and uniparental desertion of late-season young. *Auk: Ornithological Advances* 135: 427–38.

Nabe-Nielsen, J., J. Tougaard, J. Teilmann, K. Lucke, and M. C. Forchhammer. 2013. How a simple adaptive foraging strategy can lead to emergent home ranges and increased food intake. *Oikos* 122: 1307–16.

Nisbet, R., E. Muller, K. Lika, and S. Kooijman. 2000. From molecules to ecosystems through dynamic energy budget models. *Journal of Animal Ecology* 69: 913–26.

Noonan, M. J., M. A. Tucker, C. H. Fleming, T. S. Akre, S. C. Alberts, A. H. Ali, J. Altmann, P. C. Antunes, J. L. Belant, D. Beyer, N. Blaum, K. Böhning-Gaese, et al. 2019. A comprehensive analysis of autocorrelation and bias in home range estimation. *Ecological Monographs* 89: e01344.

Peacor, S. D., S. Allesina, R. L. Riolo, and T. S. Hunter. 2007. A new computational system, DOVE (Digital Organisms in a Virtual Ecosystem), to study phenotypic plasticity and its effects in food webs. *Ecological Modelling* 205: 13–28.

Piou, C., U. Berger, and V. Grimm. 2009. Proposing an information criterion for individual-based models developed in a pattern-oriented modelling framework. *Ecological Modelling* 220: 1957–67.

Platt, J. R. 1964. Strong inference. *Science* 146: 347–52.

Popper, K. 1959. *The Logic of Scientific Discovery*. Hutchinson and Company, London.

Preisser, E. L., D. I. Bolnick, and M. F. Benard. 2005. Scared to death? The effect of intimidation and consumption in predator-prey interactions. *Ecology* 86: 501–9.

Rademacher, C., C. Neuert, V. Grundmann, C. Wissel, and V. Grimm. 2004. Reconstructing spatiotemporal dynamics of Central European natural beech forests: The rule-based forest model BEFORE. *Forest Ecology and Management* 194: 349–68.

Railsback, S., D. Ayllón, U. Berger, V. Grimm, S. Lytinen, C. Sheppard, and J. Thiele. 2017. Improving execution speed of models implemented in NetLogo. *Journal of Artificial Societies and Social Simulation* 20: 3.

Railsback, S. F. 2001. Getting "results": The pattern-oriented approach to analyzing natural systems with individual-based models. *Natural Resource Modeling* 14: 465–74.

————. 2016. Why it is time to put PHABSIM out to pasture. *Fisheries* 41: 720–25.

Railsback, S. F., M. Gard, B. C. Harvey, J. L. White, and J. K. H. Zimmerman. 2013. Contrast of degraded and restored stream habitat using an individual-based salmon model. *North American Journal of Fisheries Management* 33: 384–99.

Railsback, S. F., and V. Grimm. 2019. *Agent-Based and Individual-Based Modeling: A Practical Introduction*. 2nd ed. Princeton University Press, Princeton, NJ.

Railsback, S. F., and B. C. Harvey. 2002. Analysis of habitat selection rules using an individual-based model. *Ecology* 83: 1817–30.

————. 2011. Importance of fish behavior in modelling conservation problems: Food limitation as an example. *Journal of Fish Biology* 79: 1648–62.

————. 2013. Trait-mediated trophic interactions: Is foraging theory keeping up? *Trends in Ecology & Evolution* 28: 119–25.

Railsback, S. F., B. C. Harvey, J. W. Hayse, and K. E. LaGory. 2005. Tests of theory for diel variation in salmonid feeding activity and habitat use. *Ecology* 86: 947–59.

Railsback, S. F., B. C. Harvey, S. K. Jackson, and R. H. Lamberson. 2009. InSTREAM: The individual-based stream trout research and environmental assessment model. PSW-GTR-218, USDA Forest Service, Pacific Southwest Research Station, Albany, CA.

Railsback, S. F., B. C. Harvey, S. J. Kupferberg, M. M. Lang, S. McBain, and H. H. J. Welsh. 2016. Modeling potential river management conflicts between frogs and salmonids. *Canadian Journal of Fisheries and Aquatic Sciences* 73: 773–84.

Railsback, S. F., B. C. Harvey, R. H. Lamberson, D. E. Lee, N. J. Claasen, and S. Yoshihara. 2002. Population-level analysis and validation of an individual-based cutthroat trout model. *Natural Resource Modeling* 15: 83–110.

Railsback, S. F., B. C. Harvey, and J. L. White. 2014. Facultative anadromy in salmonids: Linking habitat, individual life history decisions, and population-level consequences. *Canadian Journal of Fisheries and Aquatic Sciences* 71: 1270–78.

Railsback, S. F., and M. D. Johnson. 2011. Pattern-oriented modeling of bird foraging and pest control in coffee farms. *Ecological Modelling* 222: 3305–19.

————. 2014. Effects of land use on bird populations and pest control services on coffee farms. *Proceedings of the National Academy of Sciences* 111: 6109–14.

Railsback, S. F., R. H. Lamberson, B. C. Harvey, and W. E. Duffy. 1999. Movement rules for spatially explicit individual-based models of stream fish. *Ecological Modelling* 123: 73–89.

Railsback, S. F., H. B. Stauffer, and B. C. Harvey. 2003. What can habitat preference models tell us? Tests using a virtual trout population. *Ecological Applications* 13: 1580–94.

Riddle, M., C. M. Macal, G. Conzelmann, T. E. Combs, D. Bauer, and F. Fields. 2015. Global critical materials markets: An agent-based modeling approach. *Resources Policy* 45: 307–21.

Rykiel, E. J., Jr. 1996. Testing ecological models: The meaning of validation. *Ecological Modelling* 90: 229–44.

Santini, G., A. Ngan, M. T. Burrows, G. Chelazzi, and G. A. Williams. 2014. What drives foraging behaviour of the intertidal limpet *Cellana grata*? A quantitative test of a dynamic optimization model. *Functional Ecology* 28: 963–72.

Santini, G., A. Ngan, and G. A. Williams. 2011. Plasticity in the temporal organization of behaviour in the limpet *Cellana grata*. *Marine Biology* 158: 1377–86.

Satterthwaite, W. H., M. P. Beakes, E. M. Collins, D. R. Swank, J. E. Merz, R. G. Titus, S. M. Sogard, and M. Mangel. 2009. Steelhead life history on California's central coast: Insights from a state-dependent model. *Transactions of the American Fisheries Society* 138: 532–48.

————. 2010. State-dependent life history models in a changing (and regulated) environment: Steelhead in the California Central Valley. *Evolutionary Applications* 3: 221–43.

Schindler, D. E., R. Hilborn, B. Chasco, C. P. Boatright, T. P. Quinn, L. E. Rogers, and M. S. Webster. 2010. Population diversity and the portfolio effect in an exploited species. *Nature* 465: 609–12.

Schmitz, O. J. 2000. Combining field experiments and individual-based modeling to identify the dynamically relevant organizational scale in a field system. *Oikos* 89: 471–84.

———. 2001. From interesting details to dynamical relevance: Toward more effective use of empirical insights in theory construction. *Oikos* 94: 39–50.

———. 2010. *Resolving Ecosystem Complexity*. Princeton University Press, Princeton, NJ.

Shugart, H. H., W. Bin, F. Rico, M. Jianyong, F. Jing, Y. Xiaodong, H. Andreas, and H. A. Amanda. 2018. Gap models and their individual-based relatives in the assessment of the consequences of global change. *Environmental Research Letters* 13: 033001.

Simpkins, D. G., W. A. Hubert, C. Martinez Del Rio, and D. C. Rule. 2003. Physiological responses of juvenile rainbow trout to fasting and swimming activity: Effect of body composition and condition indices. *Transactions of the American Fisheries Society* 132: 576–89.

Stauffer, H. B. 2008. Contemporary Bayesian and frequentist statistical research methods for natural resource scientists. John Wiley & Sons, Hoboken, NJ.

Stillman, R. A., L. M. Bautista, J. C. Alonso, and J. A. Alonso. 2002. Modelling state-dependent interference in common cranes. *Journal of Animal Ecology* 71: 874–82.

Stillman, R. A., and J. D. Goss-Custard. 2010. Individual-based ecology of coastal birds. *Biological Reviews* 85: 265–76.

Stillman, R. A., J. D. Goss-Custard, A. D. West, S. Durell, R. W. G. Caldow, S. McGrorty, and R. T. Clarke. 2000. Predicting mortality in novel environments: Tests and sensitivity of a behaviour-based model. *Journal of Applied Ecology* 37: 564–88.

Stillman, R. A., S. F. Railsback, J. Giske, U. Berger, and V. Grimm. 2015. Making predictions in a changing world: The benefits of individual-based ecology. *BioScience* 65: 140–50.

Strand, E., G. Huse, and J. Giske. 2002. Artificial evolution of life history and behavior. *American Naturalist* 159: 624–44.

Thompson, J. N., and D. A. Beauchamp. 2014. Size-selective mortality of steelhead during freshwater and marine life stages related to freshwater growth in the Skagit River, Washington. *Transactions of the American Fisheries Society* 143: 910–25.

Thurfjell, H., S. Ciuti, and M. S. Boyce. 2014. Applications of step-selection functions in ecology and conservation. *Movement Ecology* 2: 4.

Tyler, J. A., and K. A. Rose. 1994. Individual variability and spatial heterogeneity in fish population models. *Reviews in Fish Biology and Fisheries* 4: 91–123.

Umeki, K. 1997. Effect of crown asymmetry on size-structure dynamics of plant populations. *Annals of Botany* 79: 631–41.

Valdovinos, F. S., R. Ramos-Jiliberto, L. Garay-Narváez, P. Urbani, and J. A. Dunne. 2010. Consequences of adaptive behaviour for the structure and dynamics of food webs. *Ecology Letters* 13: 1546–59.

Valone, T. J. 2006. Are animals capable of Bayesian updating? An empirical review. *Oikos* 112: 252–59.

Van Winkle, W., H. I. Jager, S. F. Railsback, B. D. Holcomb, T. K. Studley, and J. E. Baldrige. 1998. Individual-based model of sympatric populations of brown and rainbow trout for instream flow assessment: Model description and calibration. *Ecological Modelling* 110: 175–207.

Vincenot, C. E. 2018. How new concepts become universal scientific approaches: Insights from citation network analysis of agent-based complex systems science. *Proceedings of the Royal Society B* 285: 20172360.

Ward, J. F., R. M. Austin, and W. MacDonald. 2000. A simulation model of foraging behaviour and the effect of predation risk. *Journal of Animal Ecology* 69: 16–30.

Werner, E. E., and B. R. Anholt. 1993. Ecological consequences of the trade-off between growth and mortality rates mediated by foraging activity. *American Naturalist* 142: 242–72.

Werner, E. E., and S. D. Peacor. 2003. A review of trait-mediated indirect interactions in ecological communities. *Ecology* 84: 1083–100.

White, J. L., and B. C. Harvey. 2007. Winter feeding success of stream trout under different streamflow and turbidity conditions. *Transactions of the American Fisheries Society* 136: 1187–92.

Wilensky, U. 1999. NetLogo: http://ccl.northwestern.edu/netlogo/. Center for Connected Learning and Computer-Based Modeling, Northwestern University, Evanston, IL.

Wilson, R. P., E. Grundy, R. Massy, J. Soltis, B. Tysse, M. Holton, Y. Cai, A. Parrott, L. A. Downey, L. Qasem, and T. Butt. 2014. Wild state secrets: Ultra-sensitive measurement of micro-movement can reveal internal processes in animals. *Frontiers in Ecology and the Environment* 12: 582–87.

Wipfli, M. S., and C. V. Baxter. 2010. Linking ecosystems, food webs, and fish production: Subsidies in salmonid watersheds. *Fisheries* 35: 373–87.

Wurtsbaugh, W. A., and D. Neverman. 1988. Post-feeding thermotaxis and daily vertical migration in a larval fish. *Nature* 333: 846.

Yu, R., S. Railsback, C. Sheppard, and P. Leung. 2013. *Agent-Based Fishery Management Model of Hawaii's Longline Fisheries (FMMHLF): Model Description and Software Guide*. University of Hawaii at Manoa, School of Ocean and Earth Science and Technology, and Joint Institute for Marine and Atmospheric Research, Honolulu.

Zhivotovsky, L. A., A. Bergman, and M. W. Feldman. 1996. A model of individual adaptive behavior in a fluctuating environment. Pages 131–53 *in* R. K. Belew and M. Mitchell, editors, *Adaptive Individuals in Evolving Populations*. Addison-Wesley, Reading, MA.

Index

MONOGRAPHS IN POPULATION BIOLOGY

SIMON A. LEVIN AND HENRY S. HORN, SERIES EDITORS

Lightning Source UK Ltd.
Milton Keynes UK
UKHW021309161020
371706UK00006B/485